In Search of Giants

Dedication

In Memory of My Friends
Vladimir Markotic, who passed away in 1994
Robert Titmus, who passed away in 1997
And Barbara Butler, who passed away in 1998

In Search of Giants
Bigfoot Sasquatch Encounters

Thomas Steenburg

hancock
house

ISBN 0-88839-446-2
Copyright © 2000 Thomas Steenburg

Cataloging in Publication Data
Steenburg, Thomas N. (Thomas Nelson)
  In search of giants

  ISBN 0-88839-446-2
  1. Sasquatch. 1. Title.
  QL89.2.S2S76 2000   001.944   C99-910457-8

Printed in Canada—Kromar

Editor: Nancy Miller
Production: Annette Zacher
Cover design: Ingrid Luters

Also by Thomas Steenburg
The Sasquatch In Alberta, 1990
Sasquatch/Bigfoot, The Continuing Mystery, 1993

We acknowledge the financial support of the Government of Canada through the Book Publishing Industry Development Program for our publishing activities.

We acknowledge the assistance of the Province of British Columbia, through the British Columbia Arts Council.

Published simultaneously in Canada and the United States by

HANCOCK HOUSE PUBLISHERS LTD.
19313 Zero Avenue, Surrey, B.C. V4P 1M7

HANCOCK HOUSE PUBLISHERS
1431 Harrison Avenue, Blaine, WA 98230-5005

(604) 538-1114   Fax (604) 538-2262
(800) 938-1114   Fax (800) 983-2262
Web Site: www.hancockhouse.com   *email:* sales@hancockhouse.com

Contents

	Preface	6
Chapter 1	A Good Report	9
Chapter 2	Things Change, Things Stay the Same	23
Chapter 3	A Creature on the Road	26
Chapter 4	Joseph	48
Chapter 5	More Encounters with B.C.'s Giants	69
Chapter 6	British Columbia Statistics	110
Chapter 7	Hoaxes and the Lunatic Fringe	114
Chapter 8	Mistaken Identify & Other Errors	121
Chapter 9	The Alberta Scene	130
Chapter 10	The Crandell Campground Incident	227
Chapter 11	Alberta Statistics	248
Chapter 12	Personal Thoughts	251

Preface

Ihave been fascinated by cryptozoological (reported but uncon-firmed) creatures since I was a small boy. Sasquatch, the Loch Ness monster, Ogopogo, sea serpents and many other creatures dominated my childhood thoughts and dreams, much to the dismay of my parents at times. I remember my father telling my mother, "Don't worry he'll grow out of it." Well, as the years went by it became apparent that I wouldn't grow out of it. In fact, I wanted to pursue such creatures and find out whether or not they indeed existed or were simply folklore.

During the 1970s, the movie *The Legend of Boggy Creek* was playing in the local theater. This film was a documentary concerning repeated encounters between residents of a small Arkansas town and a large hair-covered creature that had been seen in the swamp-filled woods that surrounded the town. I was twelve years old and I must have seen that film ten times. I was hooked. Now one mysterious creature took precedence over all the others—sasquatch.

In 1977, my high school social studies teacher instructed the class to do a major study on a Canadian topic. It was up to the students to pick the subject and present what we wanted to do to the teacher for approval. You can guess which subject I chose. The teacher told me that monsters were not what she had in mind and I should choose something else. I stayed after school that day, missing my bus ride home, hoping to convince her to change her mind. I must have said something right because finally she relented.

A month later I gave my presentation to the class. My fellow students were transfixed. When I had finished my class mates applauded and my teacher sat there making notes. The next day the teacher called me to her desk and gave me my marks—110 out of 100! I was speechless. She then asked if I would give my presentation to her other classes. This I did and was well received.

After high school I enlisted with the Canadian Armed Forces as postsecondary schooling was beyond my financial capabilities. I was hoping for a posting in B.C., the traditional home of the sasquatch. I ended up in Calgary, Alberta, where my life as a sasquatch researcher began.

There, I took advantage of military wilderness survival courses and wildlife tracking. In my spare time I became a hunter. I also placed ads in papers asking anyone who believed they may have seen the creature to contact me. My phone began to ring steadily. There were many strange stories and accounts to check out. At that point I didn't have much training to evaluate the information I was given. After a dozen or so wild goose chases, I began to learn to be skeptical. I had to study the facts with a clear mind and evaluate each case carefully.

The early 1980s were a learning time for me, as well as a great adventure. In 1989 I wrote my first book *The Sasquatch in Alberta*. Publishing a book seems to give a person credibility and witnesses began to contact me without seeing the ads in the local papers. I was always talking to people, going out in the field and making contact with other sasquatch researchers. I was flooded with information and did everything I could to absorb it all, piece by piece.

In 1993 I wrote *Sasquatch/Bigfoot: The Continuing Mystery*. The book was well received and resulted in much publicity such as TV, radio and speaking engagements. All of which resulted in more witnesses coming forward to share their encounters with this elusive creature. Unfortunately, we were no closer to finding and answer to the sasquatch question than we were in the 1950s.

During a particularly cold and nasty winter in 1997–98, I decided I had enough information to write a third book. It would include many reports left out of my previous publications and new incidents that occurred after their publication. This book is the result. It is different from the others in that I included many of the witness interviews in their entirety.

I need to give my thanks to my colleagues Rene Dahinden and John Green, whose tireless work in the early years of sasquatch research inspires me. Don Keating deserves special mention too, as his work in the eastern U.S. has opened my mind to the possibility that the creature could indeed exist in the eastern North America. Danny Perez also inspires me with his enthusiasm for the hunt. Larry Lund's sense of humor when confronted with skeptics is admirable. Dr. Grover Krantz and Dr. John Bindernagel also deserve mention for their steadfast efforts, despite criticisms from their fellow academics. I would also like to acknowledge the late Barbara Butler who showed me a thing or two about tracking game in the back country. I would also like to mention the late Roger Patterson and Bob Gimlin, whose film of a sasquatch at Bluff Creek, California, continues to fascinate and survive attempts to discredit it. There are many others out there who have helped and inspired me in the last twenty years—I thank them all.

In this book you will read some of the letters and reports witnesses have sent me over the years. I have made no attempt to clean up spelling or grammar. I have also repeated what witnesses told me word for word, although I have eliminated "umms" and "ahhs" to increase readability. When people paused in their speech, I indicated this through the use of ellipses. Also, many of their drawings and sketches are included in this book.

I hope you will find this book interesting and see how strong the evidence is, even though it may be circumstantial, that we do indeed have a large unclassified creature in the wilderness areas of western Canada. If in the future it is proven that the sasquatch does not exist, then I hope I have done my part to record a great piece of North American folklore.

A Good Report

May 25, 1997, was a clear, sunny day in the beautiful Fraser Canyon of British Columbia. Mike McDonald left his home in Abbotsford early in the morning on a hunting trip in the hopes of bagging a black bear. Most of the day went by without any bear sightings, though there were plenty of signs that bears were around. Mike was sitting at the edge of a high riverbank against a tree, scanning the opposite bank with his binoculars, his rifle beside him. His camera was left about a mile away in the front of his truck. He was feeling a little tired after a whole day of trudging through thick bush and having no luck in even seeing a bear, let alone shooting one. Putting his binoculars down he leaned his head back against a tree, closed his eyes and just listened to the sound of the river rushing past and enjoyed the light breeze he now felt.

When he awoke, he didn't know how long he had been sleeping. Checking his watch he realized he had been out about thirty minutes. He again picked up his binoculars to scan the opposite riverbank. About thirty yards beyond the river, he spotted what he at first thought was a feeding brown bear.

Mike was sitting high atop a bank on the west side of the river and the bear was on the east side at a lower position so Mike was actually looking down on the creature's back, which was turned the other way. Excellent, he thought, as he put the binoculars down and brought up his rifle, placing the cross hairs on the back of the bear. His excitement was washed away in an instant and was replaced by fear as he felt the hair on the back of his head stand up. For while he was watching his prize bear through the scope of his rifle, the bear stood up and continued to stand there on two legs! Mike now knew this was no bear he was looking at. It was a creature he had heard other people talk about all his life, but never thought he would actually see one. The creature that stood before him was huge, about seven feet tall, covered from head to toe in what he described as chocolate brown hair.

Mike watched, shaking a bit, as the animal reached up with one arm to grab a branch from a tree. He then watched as the creature brought the branch down to its mouth and began to eat the leaves from the branch. When it was finished with the branch the creature turned and was now facing toward Mike and it continued to grab other branches and eat the leaves in the same manner. Mike, thinking that this creature might be a man, put his rifle down and again picked up his binoculars and continued to watch as the creature continued to feed. It hadn't seen or smelled Mike's presence, in fact he was convinced that the creature had no idea he was in the area, though through the binoculars Mike could see that this creature was very uneasy and that its eyes were continuously looking around, checking its surroundings. It seemed to be very high-strung.

Mike now had the thought, "I have to get a picture of this thing," and he started to crawl back toward the trail he came in on. When he got back to the trail he ran back to his truck and was amazed, on later reflection, about how fast he got there. He grabbed his camera and ran back to the riverbank, and then slowly crawled back to the tree he was sitting against. He again tried to find the creature with his binoculars, but it was gone. He scanned up and down the river but there was no sign of it.

A million thoughts went through his head, "Maybe it saw me leave and is over here now." Again he started to feel uneasy and started back for his truck—this time walking very slowly. Taking the safety off his rifle, he checked over his shoulder as he went. Unlike when he came raced for his camera, this time it took much longer to get to his truck, as he stopped and watched at every sound he heard coming from the trees that surrounded him. When he finally got back to the truck he wasted no time in leaving the area, raising a large dust cloud behind him until he reached the highway.

He stopped in Spuzzum to use a pay phone. He called his girlfriend in Abbotsford to tell her about what he had seen. The only other person he told was his mother. He thought about calling the police to tell them what he had seen but then decided against it knowing that he wouldn't likely be taken seriously.

About two weeks later he heard about an International Sasquatch Symposium taking place at the H. R. MacMillan Planetarium in Vancouver. Mike decided to attend and see if he could find out more about the strange creature he had seen. I was at the symposium and was scheduled to give a talk on another alleged sighting of a sasquatch at 6:30 that evening. At this time Dr. Grover Krantz, a well-known sasquatch researcher, was giving a talk on the yeren, a sasquatchlike creature reported in the mountains of southern China. As I was listening to Grover's presentation, the symposium's coordinator, Stephen Harvey, tapped me on the shoulder.

"Tom, there's this fellow outside who wants to talk to somebody about a possible sighting he had a couple of weeks ago, I think you should talk to him." Stephen then lead me out of the auditorium and introduced me to Mike McDonald. We shook hands and he told me about the strange creature he saw while out bear hunting on May 25. I found Mike to be a straightforward individual. We talked a while and Mike agreed to a full interview. During the dinner break at the symposium I set up my tape recorder and interviewed this man who seemed relieved to finally find someone who would listen to his tale and not dismiss him as a liar or someone with strange mental aberrations.

What Mike did not know at the time was the fact that he was not alone. Many honest people over the last eighteen years have told me of encounters with large, hair-covered, manlike creatures, that we in Canada call sasquatch. In the U.S.A., the most common name is bigfoot. Mike didn't know that I considered his eye witness account, if true, better than most, for in his case the creature didn't know he was there, and he was able to watch it for some time. At no time during our conversation did Mike contradict himself, nor did he hesitate in answering my questions. Following is a verbatim account of the interview.

Q: "Where did this incident occur?"

A: "About a half hour's drive north of Hope, just off the TransCanada Highway—a little town called Spuzzum, British Columbia. There's a logging road that heads east off the TransCanada Highway...actually the road splits off into quite a few smaller roads, and this was a north fork I took. I went up that north fork about five kilometers or so into the bush."

Q: "What date did this take place?"

A: "May 25th. I actually looked it up in my day timer. I keep track of pretty much everything so."

11

Q: "Was it at night or day?"

A: "It was before dinner time, around 5 o'clock or 5:15."

Q: "Describe the area in which this took place."

A: "The immediate area? It follows a power line, this logging road that I was on. There was a valley that the power line crosses through and the whole valley, including the power line had been burnt out. I would guess three or four years ago, from looking at the vegetation, things that were growing...almost no conifer trees left, they were all just the...stumps that were left standing, and quite a few deciduous trees just starting to come out in bloom. You know, new leaves and what not, so a lot of aspen and poplar trees."

Q: "What distance would you estimate you were from this thing when you saw it?"

A: "150 yards? I think that's a pretty good guess. I've hunted for years and I'm pretty good at guessing distances, like anything under 400 yards."

Q: "You were bear hunting out there?"

A: "That's right, yes."

Q: "What was your first impression of it?"

A: "I thought it was a bear. I just saw the back of it. It was squatted down and it looked identical to...a lot of bears I've seen. A lot of times when you see them, they just look like a stump with their butts to you and their heads down in the grass while they're feeding away. So that's what I thought at first, it was a bear—a chocolate brown bear.

Q: "How did you react when you saw it?"

A: "Well, like I said, I thought it was a bear till about ten seconds, watching it with binoculars, and before. Actually, I put the binoculars down, picked up my gun and then I saw it...stood up or stand up. The hair on the neck stood up...the adrenaline rush that I got, it was just amazing. Felt like I was on some kind of drug that ah...ha, ha. I don't know it almost scared me to death. I even felt nauseous at first."

Q: "What was it doing?"

A: "At first, it was squatting down. Like I said, it looked like a bear while it was squatting. But once it stood up...it stood directly up, took half a step, and grabbed a big branch of a tree, pulled the

12

branch down and started eating the leaves off the end of the branch—the buds."

Q: "How was it eating?"

A: "It was using the one hand to hold the branch down, the other hand it was using to put the food into its mouth, putting the leaves into its mouth and pulling the leaves off with its teeth, with its mouth. From what I could see at this time, it had its back to me, but within a couple of minutes it was eating off some other branches and I got a good look at it...at the front side of it. It turned around to grab the other branches that were in the area, so I got a look at the whole...right around in a circle, it turned."

Q: "Did it stand and walk upright?"

A: "Yes. Very upright."

Q: "Did you ever see it go down on all fours?"

A: "No. No, except for the first time I spotted it. It was in a crouched position. That's when I thought it was on all fours. But it was not; it was crouched or squatted."

Q: "Was it hairy?"

A: "Completely...the face was lacking some, and under the arms...I would say like where the armpits come down or where the arms would rub on its side when it walked. It was missing hair off there."

Mike used his own arms and body to demonstrate what he was talking about at this point. He then went on to explain that the hair on the creature's flanks, between the waist and the armpit, did not seem to be as thick as the rest of the body. He assumes the reason for this is the arms rubbing against the creature's sides during arm swinging as the creature walks, though he actually never observed the subject walking any more than a couple of steps during this encounter.

Q: "What color was it?"

A: "Chocolate brown– near black! But dark chocolate brown."

Q: "How tall would you estimate this thing to have been?"

A: "Somewhere around seven feet."

Q: "How heavy do you think it was?"

A: "Somewhere around 300, 350—something like that."

It should be noted that Mike is talking about pounds here, rather than kilograms.

Q: "Did you see any facial features?"

A: "Yes."

Q: "Could you describe them for me?"

A: "Sure...the head was big, that's one thing for sure. Ah, covered with fur, like I said, or hair. And under the eyes and the cheeks and nose was not covered with hair. It was pretty much open, dark skin...the teeth were yellowish or brown. I saw it eating one branch and it opened its mouth at one point, and as it was chewing I could see the teeth that were...not white, that's for sure. They were definitely yellowish or brownish...maybe they could have been white but stained. I don't know. The eyes were dark. Dark...I could see it with the binoculars, I was looking at it...even at 150 yards I...with ten-power binoculars it looked close enough, I could see its eyes were dark. But at one point I could see it look to one side and I don't know if it was reddish or pinkish on the inside of the one eye. I was focusing on his face. I mean...I could hardly take my eyes off it you know, off the face, when it was eating, when it was facing me."

Q: "Skin color?"

A: "Dark. The same color as the fur. Maybe even darker like ah...near black."

Q: "Could you describe its arms?"

A: "Like a human's only a lot longer. They were...and the fingers. I could definitely see the hands of it. The fingers were long. The arms were long, again covered with hair...bent at the elbow, but...just like ours do. But compared to the rest of its body, its arms looked thin. I don't know if that's because of the length of them or what. They didn't seem to match its body."

Q: "Could you tell if it was male or female?"

A: "No. Just from the size of it...the body shape, it would look like a well built, sturdy...you know, maybe a little bit overweight man. Just from the broadness of its shoulders and the heaviness of it."

Mike did not see any male genitalia or female mammary glands on the subject, but he does assume that the creature was a male. However, he is not positive as to the gender of the subject.

Q: "How long did you see it for?"

A: "In total? About ten minutes I would guess. Eight to ten minutes...want me to tell you what happened and why I stopped watching it?"

Q: "I'll get to that."

A: "Okay."

Q: "Did it see you?"

A: "No. Definitely not. I've hunted for years. I use a lot of camouflage gear. I even had my face painted sometimes.... I crawled into that area, quietly. Behind a ridge from where it was and I was sitting there for over a half hour before I even noticed it there, so no. It didn't even look in my direction."

Q: "Did it make any noise?"

A: "No."

Q: "Did you smell anything?"

A: "No."

Q: "Are there any physical characteristics of this creature that stand out in your mind?"

A: "The size of it. The size of its torso...it was big and thick. I don't know how to describe it, like...like a...big man with...with just a blocky body that was kind of like one shape all the way down in its torso. And the color of its hair, and its face. Those three things I think, stand out in my mind."

Q: "Did you check for footprints after it was gone?"

A: "Not that day, no. I went back a couple of days later. I was kind of...you know, the hair was still standing and I was so pumped. I was afraid to go down there once I lost sight of it. I didn't know where it went. I didn't know if it was vicious, violent, mean. I don't know, I just...no I left. I went straight to a phone booth and called home, ha, ha."

Mike, on returning to the site a couple of days later did not find any footprints in the area, nor did he find any hair on the trees from which the subject was feeding.

Q: "Did you report what you saw?"

A: "Not until now, no."

Q: "My final question, in your own words, describe what happened."

A: "The whole...thing?"

Q: "Why not?"

A: "Okay. I was...like I said at the beginning, I came across this valley that was burnt out...I crawled in along this ridge, or walked in along this ridge, very quietly in the middle of the day, maybe 4:30. Got maybe halfway or a mile into the valley, away from where I left my vehicle. I sat down and I was glassing. I was sitting down with my back against a tree. I came out of the path I was sitting on...I didn't have a good view of the valley so I came over a small crest or a knoll that brought me out into the open where I was able to glass up and down the valley, across the river, I could see the river at the bottom very clearly. I sat down and started glassing...for bears. After about a half hour I didn't see anything, I kind of put my head back against a tree I was resting against. I just sat there and closed my eyes and dozed off for a very short time. Probably half hour or so...as soon as I snapped out of it, or woke up, I started glassing again, up and down the valley. And just as my binoculars were crossing over just below me at the river, maybe thirty yards on the other side of the river, I spotted what I thought was a brown bear, the bigfoot I guess, and that's about it, from there I told you the rest.

"I watched it stand up, which just blew my mind, the adrenaline was amazing. Then I watched it eating off the...there were three or four trees in the immediate area but it was right in a clearing and it had actually to walk up to the trees away from me. I had a clear view of the whole...up to its knees, because of the size of the grass or leaves that were growing. I could see right up to the back of its knees. And with the binoculars and the scope of my gun, I watched it feeding on these branches, pulling down the branches and feeding on the leaves. After that well I...ha, ha. I was so close to shooting it, till I saw it looked very much like a man. I couldn't. I put the cross hairs right between the shoulder blades and also right on its chest when it first turned around. I couldn't shoot it, so I put the gun down. I put my binoculars back on it and I watched it feeding on these branches, and at that time I kinda snapped out of the shock I was in and I thought, 'I got to get a picture of this.' The camera was back in my truck, about a mile away.

"So I ran, I got back up on that ridge that I first walked in on and over all the dead fall and everything I ran, full blast! I mean I'm sure it was a four-minute mile. But in reality it probably took me fifteen or twenty minutes to get out to my truck, and about the same time to get back. When I got back there it was nowhere in sight. I looked up and down the valley, I glassed for another twenty minutes, but I was kind of scared at this time, I didn't know if it was...still down on the same side, gone up into the trees or up one side of the valley or the other. I didn't know, maybe it was on the same side, maybe it was investigating, maybe it saw me run, and it came...a million thoughts went through my head and ah...that was it. I thought it's going to start getting dark in about half an hour and I wanted time to get outta there. Didn't feel comfortable walking out in the dark so I even walked with the safety off and everything back to my truck. And probably walked half the way backwards, looking over my shoulders and in every direction. That's about it. I made it out to my truck. Probably 100 miles an hour all the way out to the highway again, and I immediately called my girlfriend at home and told her about what I saw, and that was it. I got home. I didn't...if I should report it to the police? I kept thinking they're going to think I'm nuts, you know. But I was willing even to take them back out to the area and show them. Then they could check it out, or use dogs or whatever they wanted. That was about it.

"Now it's two weeks later and I heard on the news on Friday that there was this symposium on bigfoot and I gotta go check this out, so here I am, ha, ha. I never expected to tell anyone, I've told two people, my girlfriend and my mother and that was it. I really, honestly thought nobody would believe me. I mean I know what I saw. I've been hunting for twenty-three years and I've seen countless animals, bears, elk, moose, everything, you name it. I've never seen anything like this, it was amazing."

After the interview Mr. McDonald took in the rest of the day's presentations at the Sasquatch Symposium. He and I talked for a long time afterwards as well. The emcee for the symposium, Daniel Perez, convinced him to go up on stage to tell the audience about his sighting. The rest of the day we made plans to go to the spot where he saw the creature.

Mike told me his vacation time was gone and he would have to check with his boss at the mill to see if he could get more time off to come with us to the site. I made arrangements to call and confirm the trip the following Monday, after the symposium was over. He drew me a map showing how to get there, in case he couldn't get more time off work. If he could, fellow researchers Daniel Perez and Barbara Butler and I would go with him to the site and hopefully at least find some evidence to back up this interesting report.

Unfortunately, Mike couldn't get away from the mill. Also Daniel Perez had to return to the U.S., so he was not able to go to the site. I phoned Mike from John Green's home in Harrison Hot Springs. John Green is a well-known researcher and author, who has written extensively on this subject over the last forty years. Mike again informed me that due to his commitments at the mill it would be impossible for him to get away.

He also didn't want to irritate his boss whom he felt had already been very good to him in allowing time off for bear-hunting season. This, of course, I could understand, having a good job myself and understanding bosses who let me take off every time somebody yells "Sasquatch!"

So on Monday, June 9, 1997, I drove up the TransCanada Highway, through the magnificently beautiful Fraser Canyon toward Spuzzum. I knew, based on the directions given to me by Mike, that the turn off was just after the Alexandra Bridge, which crosses the mighty Fraser River. The weather this day was sunny, clear and hot. Highway road crews were slowing traffic down at the bridge, which of course annoys travelers who have to stop and wait their turn to proceed. At this time I chose to pull off at a rest stop to try and get my bearings off the map. I didn't want to miss my turnoff and have to come back down in the held-up southbound traffic. As I pulled in, I heard somebody honking their horn. I turned to find Barbara Butler waiting there.

We both decided to drive down a little further in order to find the turnoff. It was only about 100 yards further down the highway. At this point though we ran into another delay. Logging trucks were using the road this day and all traffic was advised not to proceed until they were finished. So we decided to return to the rest stop and wait until it was safe to proceed. As we waited, we went over Mike's directions on how to get there. We would have to travel some distance up this logging road, which ran by the 6,485-foot Anderson River Mountain down to the Anderson River itself, where this sighting took place. As we sat there we could hear the logging trucks coming down the road which we needed to go on. After about two hours the trucks seemed to have stopped so we decided to proceed.

At the road's beginning we decided to leave Barbara's van parked there, since it wasn't four-wheel drive and we had been warned by a forestry worker that the road was in very rough shape. So she and her dog and I proceeded up the road in my 1978 Land Rover, which has never failed to get me anywhere. The road conditions were bad, but not the worst I've been on, though the ribbing caused by the logging trucks made for a very bumpy ride. After about thirty minutes the road seemed to crest a high ridge and we were suddenly on the top of a high clearing above the Anderson River. The view here of the valley was wonderful, and we could see to our left and below what looked like an old burnt out area of forest just as Mike had described. What he did not tell us about was the fork in the road at this point, so as we enjoyed the view we both wondered which way to go to get to this burnt out area. The road to the left seemed from this point to travel along the top of the valley, while the road to the right seemed to go downhill to the river.

Fortunately, at this time we spotted a pickup truck coming up the road from the right and Barbara, in the classic female way, solved our problem in such a matter that I wouldn't have even considered being the stubborn male that I am. "Let's ask these guys for directions." After a few moments of horror I relented to the logic of her suggestion. It turned out that these guys were working with a road construction crew that was improving the road further down. Mike did tell me in passing that a crew was working on the road closer to the highway when he was here in May. Now it seemed they had advanced with their work down to the river itself. They told us that the road to the right was the way to go to get to the burnt area. Even though it appeared to be heading in the wrong direction, they assured us that down by the river the road changed direction and headed north toward the burnt out area. Seeing sasquatch research written on the doors of my Rover they then began to ask questions about what we were looking for. When we informed them that there had been a recent reported sighting in this area, they seemed to be interested but they told us that they had not seen anything out of the ordinary, sighting or track-wise since they started working on the road in early May.

We then proceeded on our way; not long after we reached the river, we came to the small bridge where Mike had parked his truck on May 25. The trail he described was there and we could see that in fact it was an old road, long abandoned and overgrown with second-growth trees. It was washed out in several places by small creeks that ran down the mountain sides into the Anderson River. Many of the young trees along this old road looked as if somebody had intentionally cut them down and laid them across the road to prevent people from trying to take vehicles on it. The area fit Mike's description and we could see the power lines in the distance. Mike told me that he followed this old road/trail until he had just passed under the power lines before turning off the trail to the right passing through thick bush until sitting down by a tree on the high river bank. So putting on our packs, we followed in his footsteps to the power lines, Barbara's dog leading the way.

It was slow going. Every step involved stepping over or crawling under felled second-growth trees that covered the old road so thickly, at times we couldn't see ten feet ahead of us. One thing we did see was bear scat on the road—some old, some new. It was obvious that bears were in the area, so we kept a sharp eye out.

As we passed under the point where the power lines crossed, we decided to turn right and make our way through the dense bush to the river's bank just as Mike said he had done. We found ourselves on a fairly high, about thirty-foot, cliff overlooking the west bank of the Anderson River, just as Mike described. The east bank was below us. Even though Mike was not there with us to show the exact tree he was resting against, the area fit his description so well that I was sure we were at least within 100 yards of it one way or the other. The east bank had many patches of second-growth poplars as well as many burnt stumps from a fire that had gone through the area some years before. Of course we found no evidence of the creature seen here two weeks earlier. I would have much preferred if Mike had been there with us, that way we could have examined the exact spot where the creature had been observed, and maybe had found some hair, footprints or even some of the chewed branches the creature had been feeding on. But we had to settle for looking around in the general area, which we did, and found nothing.

We did walk along the west bank for a little ways when Barbara suddenly laughed and said, "Ah ha, evidence," reaching down in the grass and pulling up an old discarded Kokanee beer can. Before the symposium, Rene Dahinden, a colleague of ours, was filming a television commercial for Kokanee beer in the mountains north of Vancouver, in which a sasquatch steals his beer. Both Barbara and I joked about finding one of Rene's stolen cans, "Hard evidence the sasquatch was here, ha, ha." Though we did not find anything to back up Mike McDonald's claim that he watched a sasquatch on May 25, 1997, at this location, he did describe the area and it matched his description almost to the letter. Even the road crew working its way along the road was here. I have no reason, assuming that the sasquatch does indeed exist, to doubt his sighting report. Of course if the sasquatch does not exist, then this whole tale was just a well made up hoax. I happen to believe in the creature's existence, so I believe the former to be the truth. You must decide for yourself.

Barbra Butler finds old Kokanee can at the McDonald May, 1997, incident site. We joked how it must have been one of Rene Dahinden's stolen cans.

Mike McDonald, at International Sasquatch Symposium, Vancouver, B.C., 1997.

Photo: T. Steenburg.

Where Mike McDonald watched a sasquatch feeding on leaves, along the Anderson River, May 25, 1997.

Photo: T. Steenburg.

Things Change—
Things Stay the Same

Reports like the one in the previous chapter leave me puzzled and, in no small matter, frustrated! Not so much because people like McDonald come forward to report what they have seen, but because this has been going on for so long now without conclusive results.

Since the late nineteenth century, when the white man began to tame the West, there have been stories from the Native people about large manlike creatures living in the mountains that came down to steal children and women as they slept. Of course being practical and civilized, such stories were ignored as the ramblings of a primitive people who had to be changed and brought into the white man's so-called practical and civilized world. In the early years of colonizing North America, the "practical" white man seemed to forget that in the old countries men and women in their thousands were being burned alive for being witches. The dead in many places had to have stakes driven through their bodies to ensure they stayed in the grave and didn't rise to torment the living. There are some historical cases where these old fears and superstitions came to the new world with the people who settled here. Salemtown in 1692, comes to mind here. I sometimes wonder if old fears of werewolves and like creatures played a role in the early belief in North America that the only good wolf was a dead wolf—the result being the near extermination of the wolf in the United States. All this in mind, it seems astonishing that stories about large hairy people told by Native people to early explorers were dismissed as superstitious ramblings by a primitive people and were not taken seriously. Arrogance it seems played a major role during the last century when reports of strange manlike creatures came in. Today of course we no longer get reports of vampires or werewolves. Unlike these mythical creatures from the past, the sasquatch continues to be seen by more and more people every year. However, just like the arrogance of the past, most people dismiss such reports as lies and ramblings—this time of a nonsuperstitious, civilized people. The more things change, the more they stay the same.

If someone told me back in 1979, when I started looking into this mystery, that we wouldn't have an answer to the question of whether sasquatch exist at the end of the millennium, I would have told him he was nuts. Yet this seems to be the case. In the late 1990s we are really no closer to proving that this creature exists or not. I remember my friend Rene Dahinden telling me about the time all of the footprints were being found in Northern California in the late 1950s, and how he thought one would pop out from behind a tree any day. It would be grabbed and the scientific community would be turned on its ear having to declare a new species and everybody would be home in time for dinner. Well it is 1998 at the time of writing this book—not one sasquatch has been brought in and dinner is getting damn cold. Yet reports of this strange animal living in our wilderness areas do not seem to be slowing down at all. Why?

Why, if this creature is no more than folklore, do reports from people who claim to have seen it continue to come in? I can think of many strange creatures from our past that would be far more interesting and exciting to pursue than the sasquatch. But for some reason this piece of our folklore just keeps going and going. Perhaps the reason for this could be that the sasquatch is not folklore. Maybe there really is a sizable breeding population of tall, hair-covered, bipedal primates, living out their lives in the Pacific Northwest, staying mostly in remote wilderness areas and only occasionally wandering out of the back woods close to human beings. When they are close to humans they are extremely cautious, only sporadically being seen by some person who is left shocked and disbelieving of their own eyes. Later, when they have regained their senses, the witness may tell one or two people about the strange manlike creature they have encountered. Upon hearing a friend or loved one telling of such a strange encounter, most will usually express disbelief or, worse, ridicule at the person's story. The witness then decides to keep his or her strange encounter to themselves, and another sasquatch sighting will go unreported to any researcher who may be able to follow up and do on-site investigation into the incident.

However sometimes, I would estimate 10 percent of the time, the witness does report what they have seen. Despite the ridicule they have received, they do seek out a researcher to tell about their encounter with a sasquatch. These are their stories.

A Creature on the Road

On a rainy evening in September, 1987, I received a phone call from a woman named Agnes Perkins who lives in Calgary. She had seen an ad I had in the *Calgary Sun* asking for anyone who thought they may have seen a sasquatch to come forward and tell of their encounter. Mrs. Perkins, aged sixty-five, then went on to tell me that she had been driving west on the TransCanada Highway, through Rogers Pass with a friend named Charlotte White in mid-August. The two women were enjoying pleasant conversation and marveling at the splendid scenery of the Rocky Mountains. As they drove along, Agnes saw what she first thought to be a man on the right side of the road, about 800 yards ahead. As the women got closer, they both realized that the figure was not a man, but a seven-foot creature covered with black hair. When the car got closer still, the creature suddenly turned right and started to climb up the steep hillside away from the vehicle.

"It stayed on two legs the whole time!" Agnes told me over the phone. The two women did slow down as they passed the creature, but they did not stop. Agnes again turned her attention to the highway while Charlotte continued to watch as the creature started to enter thick lodgepole pine trees high up the hillside. They both lost sight of the animal as the car entered one of the highway's avalanche safety tunnels. Agnes described the creature as about seven feet tall, covered in black hair. It also traveled upright on two legs the whole forty-five seconds or so it was in sight. She was very impressed with the distance the creature put between itself and the roadside as the car passed. The two women didn't say a word to each other for about ten minutes after the sighting took place.

This is how the majority of reported sightings seem to occur. The creature walks out of its forest cover into an area, in this case a major highway, that is frequently used by people going about their everyday lives. The creature, caught by surprise, reacts as any wild animal does when it suddenly feels threatened. It moves, or in some cases it runs, for cover and the witness is left to wonder what it was he or she saw during the short time span it was in view.

I received a similar report from a fellow who did not wish for me to reveal his identity in any book, so for now I will call him Bob. Bob phoned me on May 2, 1988, to tell me about a strange animal he saw while driving north on Highway 95A toward Wycliff, B.C., at 10:30 P.M. on April 28, 1988. He was driving with a friend (he would not tell me his name) hoping to reach Wycliff before 11 P.M. Just before they came to a bridge where the highway crosses the St. Mary River, both men were shocked to see a strange upright-walking animal come out of the trees on the left side (west) of the road, and cross the highway right in front of their car. The creature did not stop but crossed the highway at a fast walking pace and disappeared into the trees on the right (east) side of the road as the car passed. The two men did not stop, but drove on to Wycliff trying to decide if they should report what they saw to the RCMP. When I asked Bob if they did report it, his reply was, "Are you kidding? No way!"

It was four days later in Calgary that Bob spotted my ad in the paper and debated with himself for about an hour whether he should call and tell me about his sighting. He described an animal that was upright walking, covered with black hair and very large. Though he had no idea just how large it was, he did think it was too tall to have been a man. We talked for a while.

"I have heard stories about these things in that area before, but I never thought that I would ever see one," he said to me. He was also surprised that the creature he saw had black hair. "I have always thought they would be light brown in color," he told me. I did ask if he would be willing to meet with me and give a full interview. At first he said no, but later he told me he would get back to me.

One day later Bob did call back and told me he had no interest in getting further involved with any investigation. He also told me he always had a reputation among his friends as a no-nonsense type and he didn't want that changed as a result of word getting out that he saw a sasquatch. I, of course, respected his wishes.

I received another report from a gentleman who, like the previous account, was driving along a road with a friend when they, too, saw an animal unlike any either man had seen before. Mr. Robert Harrison was on a fishing/camping trip with a Mr. Frank Mier on June 10, 1982. The two were driving along a dirt forestry road, about sixteen kilometers north of the town of Fernie, B.C. They were in a 1978 Chev pickup truck with four-wheel drive; both men were avid fishermen and hunters. At approximately 6 P.M., they came around a bend in the road, and were surprised to see a large hair-covered animal standing in the middle of the road; it appeared to be looking at something off to the left (south). It was standing in profile facing the left side of the road. When the truck came into view the creature turned its head to look at the oncoming truck for about two seconds then it turned and started running down the road ahead of the truck.

"For a very short time, the chase was on!" Robert told me. As the truck got closer the creature suddenly leapt off the road to the left and disappeared into the trees. The two men stopped at the spot where the creature left the road and jumped out of the truck to try to see it again, but it was gone. They did find skid marks, which they took to be the creature's impact marks as its feet touched down, nine feet down the left embankment.

"It was an incredible leap," Robert told me, "I don't think a man could leap like that without falling on his ass!" Robert later told me he thought the animal was just over six feet tall and was covered with short reddish brown hair. He also said that the face and hands of the creature didn't seem to have any hair, but the skin was a dark color, either dark gray or black. The two men did check for footprints, but other than the skid marks they found none. Later they talked about reporting what they saw to the RCMP, but decided against it. When I asked if it could have been a bear they had seen, Robert told me no. "It was upright on two legs the whole time!"

It is another report by what I would consider a reliable witness. Sometimes I get the feeling that I am going about my research the wrong way when I hear about sightings like these. When I go into the bush, I always park my Land Rover somewhere and hike into the wilderness in search of footprints and hopefully seeing a sasquatch (something which has never happened to me). But when I continue to receive reports like the ones in this chapter, I sometimes get the feeling that maybe I should just stay in my vehicle and drive along wilderness roads. Maybe I would have better luck. I have been doing this more and more over the last few years, but I've only seen a large number of deer and elk, some moose, and about eight bears, seven black and one grizzly—so far no sasquatch.

I received another roadside sighting report that occurred along the TransCanada Highway near the town of Golden during the spring of 1986. While it was not a detailed report, something about it appealed to me so I've decided to include it in this book. The main witness here was an eight-year-old boy, who I will call Jimmy (not his real name), who was with his father driving back to Calgary at night. At about 1:30 in the morning the two were approaching the town of Golden, B.C. The boy's father was very tired and so was Jimmy, so he decided to pull off the highway at a quiet place to rest and get a little sleep before carrying on. The father had just stopped the truck when his son Jimmy suddenly grabbed him, begging him to drive away. The father, concerned over his son's panic, did just that. It was a little later when they started to enter Golden that the father managed to calm the boy down enough to get an explanation from him as to what had frightened him.

"A gorilla dad, it was a gorilla. It was coming up to the truck!" The father had not seen anything, but it was obvious that something had terrified his son. A number of years later when the father picked up my first book, *The Sasquatch in Alberta*, in a library in Calgary, he decided to give me a call and talk about what his son had claimed to have seen. Jimmy, now fourteen years old at the time of my interview, told me about what he had seen that night. I found Jimmy to be a very pleasant young man who seemed a little apprehensive about talking to me. No doubt it took a little encouragement from his father to talk to me.

Q: "How old are you now Jimmy?"

Jimmy: "Fourteen."

Q: "How old were you when this happened?"

Jimmy: "Somewhere around seven or eight, somewhere around there."

Father: "He was eight."

Q: "This happened during the spring of 1986?"

Jimmy: "Yep."

Q: "Just tell me what happened."

Jimmy: "Okay. My dad had pulled over. And I looked around to see where we were. I looked out the window and I saw a long shadow and it was moving, and I just jumped on my dad, and I started crying, because I saw something moving, and my dad just left."

Q: "You thought it was a gorilla?"

Jimmy: "Yeah, that's what I thought."

Father: "That's what he called it, a gorilla."

Jimmy: "It didn't look like a bear."

Father: "What's really strange, is you have this eight-year-old boy, why would he call it a gorilla? In the middle of you know British Columbia, gorilla? That's what I thought. Why would he call it a gorilla? That's what I said to myself. It must have been a bear. That's why I left."

Q: "Does anything about the animal you saw stick out in your mind Jimmy?"

Jimmy: "It was on two legs. It was walking around I think. And it was leaning against our truck I think."

Father: "Like he...he was sitting like this right, and he jumped like that, you know, he jumped on me saying 'Let's go eh! I'm scared, I'm not staying here!' Well it's been a long time you know, and I don't think he can remember much now. But whatever he saw, it must have been very close to our truck because it scared the hell out of him."

Q: "Were you the only vehicle parked in this area at the time?"

Father: "Yeah, nobody else was there, and it happened as soon as I pulled over you know? This thing could have been there and we just happened to have approached it, without seeing it. I pulled over and just turned the motor off, and shut off the lights when he jumped on me, and off we go again."

Q: "In your own mind Jimmy, what do you think it was?"

Jimmy: "It didn't look like a bear. It was an animal. I don't know what kind but some kind of animal. But it couldn't have been a bear, it was upright. It could have been...it's hard to say."

Q: "Do you think it was a person?"

Jimmy: "Oh no, I don't think so."

Father: "I thought that later."

Jimmy: "I saw a person?"

Father: "Yeah son, later I thought maybe you saw a hitchhiker because this gorilla thing was you know, humanlike right. So later I thought hitchhiker, but I don't know. I didn't see nothing, so...all I saw was my frightened kid you know."

Q: "He was frightened by it?"

Father: "Oh, he was scared, ha, ha, no doubt about that."

Jimmy: "I just wanted to get out of there."

Father: "That's right he was...'Let's go, let's go,' ha ha."

Jimmy: "I thought it might break the window because it was really close. So I thought..."

Q: "It was approaching the truck?"

Father: "It must have been, whatever it was, it must have been very close because he jumped right on me."

Q: "But you didn't see it?"

Father: "I didn't see nothing, no."

Q: "It was dark?"

Father: "No lights in that area, that's why I pulled off there. That's why I picked the place, because it was so nice and quiet, and I thought this is a good place where we can have a good rest. It is before you hit Golden so there was nothing there. Just this nice area where you can pull over eh, quite a bit off the road. So I don't know."

I found it interesting that Jimmy did not come right out and say it was a sasquatch he saw that night, he just says it was some strange animal. His father just shrugs his shoulders and says, "Whatever it was, it scared the hell out of him!"

As can be seen from reports like these, if sasquatch do indeed exist, encounters with people who happen to be driving by in their car at the right spot at the right time, one would almost think that someone should have hit and killed one by now. It should be remembered that even though the people in this chapter did see a sasquatch at a particular point along a highway, we must consider the fact that the vehicles in question were probably only one in more than 1,000 vehicles that passed by the spot where the creature was seen in any twenty-four-hour period. I'm sure the number of vehicles that pass a particular spot on a dirt forestry road are quite a bit lower than on a major highway, but the same would hold true.

The following roadside encounter came to my attention while I was having a drink in the Hope Hotel in August, 1986. I had my maps laid out on my table trying to decide which way I was going to proceed the next day. A Native fellow at the next table asked what I was doing, so I told him. He seemed quite interested and asked if he could join me at my table. I said sure, introduced myself and we shook hands. He told me his name was Sonny and he then began to ask me questions about my search for sasquatch.

At first I thought he didn't believe in the creature's existence, then he suddenly made the statement, "My brother-in-law saw one back in the early 1970s on the highway by Ruby Creek." Well now my interest was up so I asked him for details. "My brother-in-law was on Highway 7 and apparently he saw one on the side of the road."

"Maybe he saw a bear," I said.

"That's what I said when it happened. But after all this time, he still says he saw a sasquatch. I didn't believe him at first. Nobody did. But after all this time, and all the ridicule he went through from his friends and family, he's never changed his story!"

I learned from Sonny that his brother-in-law was a fellow named Ralph Bobb, who lived just off the TransCanada Highway near Spuzzum. Sonny's sister Jennifer was also in the car when the sighting occurred. Sonny then gave me the phone number and suggested I give them a call the next day. "He might be happy to talk to somebody who will listen to him without laughing for a change."

The next morning I gave the Bobbs a phone call. I talked to Sonny's sister, as Ralph was out for the morning and wouldn't be back till the afternoon. However, she did invite me over to do an interview. I thought this would be a good chance to interview her about the incident without her husband there, then I could compare the two versions for discrepancies. I found Jennifer to be a pleasant woman who made me a cup of coffee and invited me to sit in the livingroom and wait for her husband to return. She suggested we watch TV until he came home. I suggested that I interview her in the meantime to which she agreed. From her I found out the date this alleged sighting took place—late August, 1973. She could not recall the exact date though, neither could Ralph when I talked to him later that afternoon.

Jennifer's version:

"Ralph, the kids and I were at a triple feature at the Chilliwack drive-in. We were coming home along Highway 7, past Kitty Corner, towards the Ruby Creek bridge. It was late, between 1 A.M. and 1:30, on a beautiful clear night. We were approaching the Ruby Creek Bridge, when I leaned over the front seat to cover the kids, who were sleeping in the back of the station wagon. Suddenly, I heard my husband cry out, 'Holy shit!', and at the same time he hit the brakes and brought the station wagon to a screeching halt. I didn't say a word. I just looked out the windshield to see the deer or car Ralph had braked to miss.

"'Did you see that!?' he asked.

"'No, I was covering the kids, see what?' He did not reply, he just drove up the highway a short distance and did a U-turn. My husband turned white as a ghost as we drove back very slowly.

"'Did you see it?!?' He asked again.

"'No, I was covering the kids.'

"'There was something there. It was big.'

"My husband is a logger, a hunter, and spends a lot of time in the bush, and in doing that all his life, he is very good at recognizing game. He is not easily frightened.

"'What did you see?'

"'I don't know, but it was big,' he said.

"'Maybe it was a bear?' He looked around for a few minutes with the car headlights. I just reached around and locked all the car doors. Later as we began to head home I again asked him what he saw.

"'The only thing I can think of is that it was a sasquatch.'

"'Maybe it was a bear standing up,' I suggested.

"'Not that big. Besides I know a bear when I see one.'

"The next day, we went back and looked the area over. We didn't find any tracks, the area was too rocky and hard."

Jennifer never saw what it was that had shaken her husband up so badly that night. She just remembers that he was as white as a ghost and she spent the rest of the trip home trying to calm the children down after they had suddenly been awakened by her husband slamming on the brakes. When Ralph came home an hour or so later he seemed annoyed a little bit at having this stranger in his house with his wife. But after introductions and finding out the reason for my unexpected visit, he told his version of events that night back in 1973.

Ralph's version:

"I have been hunting all my life, and I have seen a lot of strange things. But nothing could compare with the sasquatch I saw that night on the side of the road. I had heard a lot of stories about the sasquatch living where I do, but I never thought I'd ever see one. I always thought, if I ever saw one of these things, it would be in the bush on a hunting trip or something like that. Not on the side of Highway 7 watching my car go by!

"Like my wife said, we were driving home from the Chilliwack drive-in late at night. I'd say between 1 A.M. and 1:30, along Highway 7. I had an uneasy feeling like I get when I am out hunting sometimes. Hunter's feeling. I can't explain it, the hair on the back of your neck stands up, you know. Anyway, we were approaching Ruby Creek Bridge. The kids were asleep in the back of the station wagon, Jennifer was covering them with a sleeping bag. I was just driving along, when I saw this huge creature standing on the left side of the road, looking at us. It didn't do anything, it just stood there and watched the car go by. I don't remember what I said, but I slammed on the brakes and brought the car to a halt. Lucky for me it was late at night and there were no other cars on the road or I probably would have been rear ended for sure. Jennifer didn't say anything, she just sat there and wondered what the hell I was doing. 'Did you see it?' I said.

"'No I was covering the kids.'

"I did a U-turn and went back to where the creature was standing, but it was gone. I searched the area with the car lights, I did not have a flashlight with me, but I found nothing. The creature probably moved back into the trees the moment I hit the brakes. I don't believe it crossed the highway after I drove by, but then again it might have. It was so big it probably could have crossed the highway in two or three strides. The next day Jennifer and I went back to the area to look for tracks, but we didn't find any. We looked everywhere, but the area was too rocky for footprints to have been

Ruby Creek Bridge. It was just before this spot, where Ralph and Jennifer Bobb encountered a creature which Ralph insists had to be a sasquatch.

Ralph and Jennifer Bobb, in front of their home when I interviewed them in 1986.

left behind. Nobody believed me when I told them about what I saw. I really don't care what people say. I know I saw a sasquatch that night."

After my interview with the Bobbs, I stayed with them in their home for the rest of the day, answering as many of their questions about the sasquatch mystery as I could. It was during this pleasant afternoon that I was told that Ralph is a distant relative of the deceased Chapman family of the famous Ruby Creek incident of October, 1941. Anyone reading this book who knows anything about the history of the sasquatch in Canada will recall this classic incident when Mrs. Jennie Chapman ran from the family cabin with her children when a creature fitting the description of a sasquatch approached the home. Later the same week when the creature came back and made noises outside the Chapman's cabin at night, the family abandoned their home never to return. The cabin, left to the elements eventually rotted away, and today there is no trace of it left. Enough has been written on this incident in other books so I will not go into detail now. But I did find it interesting that Ralph happens to be a distant relative of the Chapmans, though he didn't seem to know much detail about the story itself. As I was getting in my truck to leave, Ralph said to me, "You know Mr. Steenburg, in all the years since I saw a sasquatch, you are the first person outside the family that ever believed me."

Another B.C. highway report came my way in 1989, when I received a phone call from a Mr. Randell Colclaugh, who told me of a strange creature he saw while driving on Highway 97, just outside the town of Quesnel. Randell's job required him to drive along the highway in the early hours of the morning, delivering newspapers at various locations between the towns of Cache Creek and Prince George. He had driven this route many times without seeing anything stranger than a deer on the road in front of him. This night however, he would see something that would haunt him for years. It happened one hot summer night in 1978, Randell this time had a female coworker along for help. Randell was visiting friends in Calgary when he heard me talking on a CBC radio talk show discussing the sasquatch mystery. He decided to give me a call. After some discussion he agreed to come to my home for a full interview.

Q: "Would you state your full name please?"

A: "Randell Ford Colclaugh."

Q: "Where did this incident occur?"

A: "Just outside of Quesnel, on the highway, about 3 o'clock in the morning."

Q: "Describe the area in which it took place."

A: "It is a forest area, very dense lodgepole pines, very thick vegetation."

Q: "What distance would you estimate you were from this thing when you saw it?"

A: "I was probably 200 feet."

Q: "What was your first reaction?"

A: "Surprise."

Q: "How about the woman who was with you?"

A: "She was surprised too."

Q: "What was it doing?"

A: "I don't know. Bending down in the ditch I guess."

Q: "Did it stand and walk on two legs?"

A: "Yes it did."

Q: "Did you ever see it go down on all fours?"

A: "No. Oh well, when it climbed up to the bank I think it used its hands to grab the ground."

Q: "Was it covered in hair?"

A: "Yes."

Q: "What color was it?"

A: "Brownish black."

Q: "How tall would you estimate this thing to have been?"

A: "About eight or nine feet tall."

Q: "What would you have estimated its weight to have been?"

A: "Oh...about...seven to eight hundred pounds."

Q: "Did you see any facial features?"

A: "Not really. But the face was like that of an ape, and it had strange ears."

Q: "Describe the ears?"

A: "The ears were...like a large lynxtype ears, the lynx cat."

Q: "Could you describe the arms?"

A: "Apelike, hairy."

Q: "Could you tell if it was male or female?"

A: "No."

Q: "How long did you see this thing for?"

A: "Forty-five to fifty seconds."

Q: "Did it make any noise?"

A: "No."

Q: "Did it see you?"

A: "Yes...well it saw my van. I don't know if it saw me or not."

Q: "And when it saw your van, what did it do?"

A: "It ran up the bank. I would say very frightened."

Q: "When it moved, was it walking or running?"

A: "It was running."

Q: "Did you smell anything before, during or after this sighting?"

A: "No."

Q: "After it was gone did you check for footprints?"

A: "No I did not."

Q: "Did you report this to the RCMP?"

A: "No."

Q: "Did you report this to anyone?"

A: "Just family and a few friends."

Q: "In your own words, describe what happened."

A: "Well as I was saying before, I used to drive a newspaper van from Cache Creek to Prince George, four days a week. And it involved driving all night long, dropping off newspapers in the towns as I went along. One night as I was driving outside of Quesnel, heading north to Prince George on the old highway, the highway has been changed now, there was a large curve on the road. And as I came around this curve my headlights filled the ditch and I startled this sasquatch creature. It was obviously scared and ran up the bank, thirty feet at an incredible rate of speed. It was gone before we knew it. Incredible strides. I slowed down and the woman I was with said, 'Do you want to stop?' I said, 'Are you crazy?' I just kept on driving. I didn't stop on the way back either, because I was scared about stopping in that area. I did stop there on another trip, about a week later, to check for tracks, but I did not find any at the time."

Q: "What was the woman's name?"

A: "I don't remember now. I just worked with her, 'eh."

Q: "Does she live in B.C.?"

A: "I think her name is Millward. She is probably married by now."

Randell's description of the animal does fit the common description of a sasquatch with one exception—the large lynxtype ears he describes. Most reports from witnesses describe very small ears if any are noticed at all. If I could locate the woman he was with that night maybe she could clarify this strange description of large ears. But all attempts to locate the woman have failed. So I have to conclude that either Randell and a woman named Millward did indeed see a sasquatch that night in 1978 or they saw some other animal with large ears they mistook for a sasquatch. Running upright does seem to rule this possibility out though. There is another option to explain this story—it never happened, and the large ears he described are the result of him failing to do his homework before contacting me. I didn't find him to be the sort of person to seek attention by making up a false sasquatch story. But then again the ears do not fit.

I was contacted by a Mr. Voytek Tertell, of Calgary, Alberta, who wanted to tell about a strange creature he saw late at night while driving east through the Rogers Pass on September 18, 1991. At about 1 A.M. Voytek, his wife and a friend had just driven through Rogers Pass and it seemed they had the whole highway to themselves. The two women, his wife Elisabeth and her friend Nancy Monsery, were asleep in the back seat. Voytek was listening to music on the car radio. Suddenly he saw on the right-hand side (south) of the road a large creature, with brown hair which he at first took to be a grizzly bear. The creature was walking along the roadside, also heading east, so the animal had its back to his car. As they approached the animal, it veered off to the right from the roadside toward the trees. As the car passed the animal, Voytek realized it was walking on two legs the whole time it was in sight. He did not stop but carried on. When Voytek contacted me to report what he had seen he did agree to a full interview, and invited me to his home.

Q: "Would you state your full name please."

A: "Voytek Tertill."

Q: "Your age?"

A: "Thirty-eight."

Q: "Where did this incident occur?"

A: "Just east of Rogers Pass, B.C."

Q: "You were still in British Columbia?"

A: "I'm not quite sure how the border goes, I think I was still in B.C. The border runs closer to the town of Field. Field is still in B.C.?"

Q: "Yes."

A: "I was well into B.C. then."

Q: "What date did this take place?"

A: "September 18, 1991."

Q: "Was it at night or day?"

A: "At night."

Q: "What time was it?"

A: "I think it was around 1 o'clock in the morning."

Q: "Describe the area in which this took place."

A: "Well, it was along the highway. I was traveling east, and this animal was walking along the right-hand side of the road. It was walking eastward too. So I saw the back of it."

Q: "What distance were you from the animal when you saw it?"

A: "When I first saw it? Oh, I don't know...100 meters (300 feet)?"

Q: "You drove past it did you?"

A: "Yeah."

Q: "What was your first impression when you saw it?"

A: "I thought it was a bear. I was tempted to hit the brakes and wake everybody and say, 'Hey look at that animal.'"

Voytek did not hit the brakes though. I wish he would have, then maybe we would have three witnesses here rather than one.

Q: "You weren't alone?"

A: "No. I was with my wife and a friend of ours."

Q: "Did anyone else see it?"

A: "No. They were asleep."

Q: "What was it doing?"

A: "It was walking on its hind legs. That's about it."

Q: "Was it walking along the road?"

A: "Yeah, walking close to the highway and as I was passing it, it seemed to wander off toward the forest."

Q: "Did it stand and walk on two legs?"

A: "That's right, yeah."

Q: "Did it ever go down on all fours?"

A: "No."

Q: "Was it hairy?"

A: "Yes, it was hairy all right."

Q: "What color was it?"

A: "Brown."

Q: "How tall would you estimate it was?"

A: "Well...it was quite a bit taller than me, 200 centimeters or seven feet. It was fairly tall. When I think of it, it was fairly slender too. It wasn't like a bear. It's an afterthought. A bear would be, you know, fatter? This thing was fairly slim."

Q: "What would you estimate its weight to have been?"

A: "Oh that's a tough one. I don't know. I can only do it in comparison to myself. I weigh 90 kilograms, this thing would oh...I don't know, 120 kilograms (300 pounds), something like that."

Q: "Did you notice any facial features?"

A: "No."

Q: "Could you describe the arms?"

A: "I really did not pay that much attention then, but when I thought about it later, they did seem a little longer than a man's."

Q: "Below the knee?"

A: "Not below the knee, but they were longer all right."

Q: "Could you tell if it was male or female?"

A: "No."

Q: "For how long did you see this creature?"

A: "Maybe five seconds."

Q: "Did it make any noise?"

A: "That's hard to say. I was listening to the music on the radio and I was in the car."

Q: "Did it see you?"

A: "I don't think so. Maybe it didn't see me, but it heard me. My car that is. As I was passing it, it sort of wandered off to the right towards the woods, so the car scared it, or the lights, whatever."

Q: "Did you smell anything?"

A: "No."

Q: "Did you stop to try and see it again, or check for footprints?"

A: "No...no way!"

Q: "Did you report what you saw to the RCMP?"

A: "No."

Q: "Did you report it to anyone?"

A: "Well to you after a year or so. It just happened I saw your book, (Sasquatch in Alberta), I thought you might like to know."

Q: "My final question, describe what happened, in your own words."

A: "I was driving on the TransCanada Highway, towards Calgary, traveling east, just past the Rogers Pass. It was fairly close to the Rogers Pass. We had just gone through it. All of a sudden I saw it in the headlights, what I thought was a grizzly bear. I was going about 110 kilometers an hour (75 mph) so I didn't see it for a very long time. I'm positive it was brown. I'm positive it was walking upright. Well built, but not fat, slim sort of thing and it was traveling in the same direction, so I didn't see any facial features. Just as I was passing it, it kind of wandered off to the right towards the woods. That's about it."

Q: "After you passed it, did you stop or anything?"

A: "No."

Q: "What were the names of the two women with you?"

A: "My wife, Elisabeth, and her friend, Nancy Monsery."

Q: "They didn't see it?"

A: "That's right, they were asleep."

Another report of a strange animal on two legs along a major highway—could all these people be mistaken? I now never go on the road without having my camera handy, even if I'm not going out into the field to do research. If I'm going anywhere that involves driving though forest regions I always have a camera ready, just in case.

I will close this chapter with one more possible roadside sighting that took place in May of 1997. The witnesses were a couple from Somerset, England, who were on vacation taking a driving tour of British Columbia and western Alberta. The couple were driving along the TransCanada Highway, west of Hope, B.C., when they thought they might have caught a glimpse of a sasquatch in a field off to the left (north). Just as in many of the incidents in this chapter, they only saw what they thought might have been a sasquatch for a few seconds as they drove by at high speed. Also, like many of the incidents in this chapter, they were very confused by what they saw. Unlike the other reports in this chapter, this time the witnesses had a video camera with them!

Mr. Wayne Oliver and Miss Julie Ellif, both of Somerset, England, were looking forward to their road trip across B.C. They would be video taping from their car almost the whole time. Judging from the tape they sent, it was raining quite hard in Vancouver when they started out. But a little later the clouds broke and the couple seemed to enjoy the high mountains topped off with clouds as they drove along the TransCanada Highway in the lower Fraser Valley. Wayne was driving the car while his girlfriend was video taping along the way. Fortunately just a few seconds before this incident took place, we see on the video a sign telling motorists that the Peters Road turnoff is just ahead. This gave me the exact location of the field this alleged sasquatch was walking in. Neither Wayne nor Julie, being first-time visitors to Canada, had any idea where they were. When they first contacted me all they could tell me was the next town they came to was Hope, B.C. Actually they were closer to the towns of Chilliwack and Agassiz, but they did not know that at the time. It was later when they had returned to England and viewed the tape of their trip that they thought they might have something of interest.

Wayne, having a keen interest in the sasquatch mystery, then went looking for books on the subject to find out more about what they might have seen. He came across my second book, *Sasquatch: Bigfoot The Continuing Mystery*, and decided to give me a long distance call to tell about the video. I was just getting ready to leave for the International Sasquatch Symposium, in Vancouver when I talked on the phone with him. He agreed to send me the tape so I could have a look at it. He didn't say he had a video of a sasquatch, he simply said he had something on tape, he didn't know what it was, could I have the tape looked at, and get back to him with the results. This I agreed to do. The tape arrived in the mail about a week after I had returned from Vancouver. What I saw did not look promising. Essentially, you see the road sign telling motorists that Peters Road turnoff is ahead, then a few moments later you hear Wayne's voice say, "What's that?" Julie then puts the camera on him and you can see him pointing to something while he's driving. In the field beyond his pointing finger you can see what appears to be a large erect object in the field some distance away, north of the highway. The image is blocked repeatedly by westbound traffic, and a road side batch of trees. The object is only in view for about five seconds. During this five seconds you hear Julie say, "Sasquatch."

"What's that?" Wayne replies.

"Sasquatch!" she says again, in a raised voice. By this time the car has passed the open field and only forest is visible. The two did not stop or turn around but continued on to Hope, and then carried on with the rest of their trip. Why they did not stop, I will never understand. Perhaps they did not realize they may have had something until they viewed the tape, or maybe the traffic on that part of the highway deterred them from stopping. The latter I can understand for that part of the highway is very busy with traffic and pulling over and getting out of your car can be very hazardous. Wayne, on reflection, questioned his own actions at the time in the accompanying letter he sent with the video.

Mr. Steenburg:
    I've eventually copied the sighting and I've also included the original. If I could have the original back when you're finished I'd be grateful. I hope you find something on this tape as I'm sure I saw something strange. Why I didn't turn around I'll never know. If you could contact me after you've viewed it, I'll be grateful.

Thank you,
W. J. Oliver.

The video Wayne sent me was of poor quality due to the fact he played the original on his TV, and then filmed it again on another tape. Fortunately he also sent the original along. The tape is 8-mm Pal cassette, which is commonly used in European home video cameras, but not used in North America. So I had to get the original transferred to VHS. Once this was done I then took the tape to the

University of Calgary, photo/graphics department, who in turn made stills from the video. Supervisor Blair Pinder was most helpful, having become interested himself as to what might be on this tape. He made blow ups of the stills on his computer. Unfortunately, video has a major drawback that old 8-mm home movie cameras do not. You cannot freeze frame them without losing clarity. Also when you blow up frozen video images more clarity is lost. So the stills from the video were nothing more than shapeless blobs. In the meantime, I had sent the original video tape back to England. What was determined by the photo/graphics department was this was not some stationary object, tree stump, rock or whatever. It is a video of a large, upright moving object, which is dark colored. So I have to conclude that Wayne and Julie either videotaped a large man in the field that day or they really did tape a sasquatch.

Blair Pinder suggested I try to get the original tape back again so more work could be done. This I agreed to do. I sent Wayne a letter with what we had concluded and asked if he could again send me the original 8-mm Pal tape. About mid-August the original tape was back in my hands with the following letter.

Dear Thomas

I've just received the color photos and the letter you sent which I've found very interesting. To me it just seems too large and too black to be a man, as everything around it seems that the colours are so sharp and clear. I can't give you much more information as it was now nearly 4 months ago when the incident happened. I remember seeing something unusual move slightly, (I guess it was coming out of the trees) toward the water area. Maybe it was going to wash or take a drink? My first reaction was that I couldn't believe what I had seen. It just seemed too weird to see something I had been very interested in for many years, the first time I actually come to Canada. I remember that we were going along the road and my ex-girlfriend was trying to capture me singing on film, but through embarrassment I looked the other way out of the window. It was then out of the corner of my eye I saw it. It just looked so unusual and I know I wouldn't have reacted if it was a man, as we saw so many men in fields along the trip. I remember thinking 'How could I see one on my first trip'. And instead of turning around I thought I'd wait until I'd get home to check out what I'd captured. I regret not turning around so much as you can imagine.

Thank you for all the work and research you've been doing and hopefully it will pay off. I'm enclosing the original video again so hopefully it may help. Please be sure to contact me if anything originates from the original shot, or you know for sure what it is.

Many thanks,
Wayne.

By the end of August, the University of Calgary photo/graphics had examined the original video again, and unfortunately the image in question was not in view long enough for any firm conclusions to be made. Also, frozen stills from the original again lost too much clarity during the enlargement process. So we, at the time of writing, are left with the same conclusions. Wayne and Julie did indeed video tape a large, moving, upright object. Whether or not it was a sasquatch or just a large man cannot be determined at this time. I did send copies of my stills to John Green in Harrison Hot Springs, B.C. Since he was close to where this incident took place, he made a trip out to the site and again confirmed that no stationary object was in the field where this video was shot. In John's letter to me he writes:

Dear Tom:

I haven't had much in the way of spare time lately, but today I did go looking for the Oliver video site. It is not as simple a matter as it seemed. Peters road is no problem, it is actually closer to Agassiz than to Hope, and the configuration of the mountain skyline gets you to a very short stretch of road before it changes again from what the pictures show. There is however, a very obvious sloping short post apparently close to the west bound lanes of the road in one of the pictures. That would have served to pin down the location exactly, but we could not find anything remotely resembling it. Because of the steady traffic you cannot walk around in the traffic lanes, so of course you can't get the exact angles involved in getting the video shots, but I am sure we couldn't have been more than 50 or 100 feet out of position. That being so, the figure is on the beach across a slough from the road. There is also the main line of the C.N.R. between the road and the slough, and the river is just a short distance farther away. There was nothing there to account for the figure in the video while we were studying the area, but when we stopped there on the way back there was, a human standing on the beach at the far side of the slough. That was at dusk, and the figure just looked black. Presumably it would be a person fishing. At that location I would say the odds would be 1000 to 1 that any upright walking figure would be a person rather than a Sasquatch, unless they had a clear enough look at it to be sure it could not have been a human.

John Green
October 6, 1997.

When I last talked to Wayne Oliver on the phone and told him of John Green's findings he again expressed the opinion that he and his ex-girlfriend did not see a human but something else. So again, this sighting is inconclusive. What do you think?

Stills from the Wayne Oliver video taken on May 5 or 6, 1997.

Video: Wayne Oliver.

Joseph

In May, 1996, I was asked to attend another forum on the sasquatch mystery. The forum was to be held in Harrison Hot Springs, British Columbia, in conjunction with the town's annual Sasquatch Daze celebrations. The Harrison Lake area is well regarded by those of us who have an interest in this mystery as the traditional home of the sasquatch in Canada. Harrison Hot Springs on the south shore of the lake has been seen as the gateway to sasquatch country, so to speak, for about the last forty-five years.

Many of the town's residents quite enjoy this unofficial title bestowed on their town, while others wish the whole sasquatch thing would just go away. There have been periods in the past where some of the town's residents have made serious attempts to erase most traces of the sasquatch from their community. I recall an old news clipping in my files from *The Province*, a Vancouver newspaper, dated Tuesday, May 24, 1966, in which some residents were successful in having an old illuminated sign that read, Welcome to Harrison Hot Springs, Land of the Sasquatch, removed from the road at the town's entrance. The part of the sign with the greeting was held by a smiling sasquatch with a top hat on its head and its other hand stuck out in the classic hitching-a-ride pose. Some residents at the time thought their resort community was outgrowing the image of being the gateway to sasquatch country. The old sign was replaced with a tasteful overhang sign held in place at both sides with Native totems. This sign stayed in place for the next twenty-nine years until it too was removed in 1995, some say due to the fact that it was rotting and was in danger of falling down on a car. Others grumble it was removed to allow high logging trucks to drive through town on their way up the lakeside road. In 1996, a new third sign was erected at the town's entrance. Residents who were against the first sign in 1966, if they were still around, I'm sure were unhappy to see the sign again featuring a picture of a walking sasquatch welcoming guests. Visitors to this wonderful resort community don't really need a sign at the town's entrance to realize the hairy biped plays a major role in the town's culture. The drive in passes

Left: The sign which used to greet visitors to Harrison Hot Springs, B.C. In 1966, several of the residents lobbied successfully to have the sign removed, trying to distance their community from the sasquatch legend.

Photo: The Province, *May, 1966.*

Below: Old legends die hard though. This new sign was erected in 1995. Once again the sasquatch greets visitors to Harrison Hot Springs.

Photo: T. Steenburg.

Bigfoot Campgrounds, sasquatch souvenir stores and other signs directing campers to a provincial park, named Sasquatch Park. (It seems to me the town finally gave up trying to distance themselves from the overgrown creature that many say wanders the mountains of Harrison Lake.)

In 1993, the town started once again to promote Sasquatch Daze, which used to be a Native celebration held in the area before World War II. In the new Sasquatch Daze, visitors were invited to participate in many activities, like sasquatch races (teams racing with large wooden feet tied to their shoes), sasquatch hikes, Art on the Lawn (local artisans displaying their work), Sasquatch Market (a flea market), arts and craft sales, Harper's Bazaar (local craft sale) and, of course, the sasquatch parade. It was all great fun, and became a yearly event from 1993 to 1996. Due to money matters, 1996 saw the last Sasquatch Daze. I hope the tradition will start up again, but at the time of writing I'm told there are no such plans.

Stephen Harvey, who was one of the main planners of Sasquatch Daze and who now organizes the Vancouver sasquatch symposiums, had the idea of a serious forum to take place in the town during the celebrations. Many of the most well-known researchers were invited to give talks and exchange information. The forum was also open to the public. I was asked to give a presentation at the first forum in 1993, and I gave presentations at the forums in the following years.

It was during the Sasquatch Daze of 1996 that I first met Joseph Verhovany. I was staying in town at Karas Trailer Court, it was the last day of the forum and I was making plans to spend the next week in the mountains around Harrison Lake when I heard a knock on my trailer door. Joseph had seen my Land Rover with its sasquatch research signs on the doors parked outside my trailer. (I debated about putting signs on my vehicle for years and I finally decided to do it—a decision that has paid off with several reports from people who have seen me drive by.) He decided to finally tell somebody about two encounters he had with creatures he believes were sasquatch.

The first occurred in July of 1988. The second occurred in March of 1993. I found him a nice, easygoing fellow who spoke with a thick accent. As we were talking, colleague and fellow researcher Daniel Perez arrived. We both sat down with Joseph to have a detailed interview about his encounters.

Steenburg: "Where exactly did this incident occur?"

A: "Exactly by the Silver River and Harrison Lake mouth. That is where the Silver River flows into Harrison Lake."

Author's note: Joseph, as I said before, speaks with a thick accent. I do not try to clean up the grammar of people I interview. I feel the reader gets a better idea of the witness' personality if I write exactly what they said, word for word.

Steenburg: "What date did this take place?"

A: "July 5th, 1988."

Steenburg: "Was it night or day?"

A: "Day."

Steenburg: "What time of day?"

A: "It was about four or five in the afternoon."

Steenburg: "What was the weather like?"

A: "Sunny."

Steenburg: "Describe the area in which this took place."

A: "This area is wild berry growth, right on the edge of the Silver River."

Steenburg: "What distance would you estimate were you from this thing when you first saw it?"

A: "About 100 feet...100, 150, something like that."

Steenburg: "What was your first reaction when you saw it?"

A: "I thought I seeing a grizzly bear."

Steenburg: "What was it doing?"

A: "Picking berries."

Steenburg: "Did it stand and walk on two legs?"

A: "Two legged."

Steenburg: "Did it ever go down on all fours?"

A: "Never see it go down."

Steenburg: "Was it covered with hair?"

A: "Yes sir."

Steenburg: "How tall do you estimate this thing to have been?"

A: "Way taller than me, over six feet."

Steenburg: "What would you estimate its weight to have been?"

A: "300, 350 pounds, minimum."

Steenburg: "Did you see any facial features?"

A: "Yes."

Steenburg: "Could you describe them?"

A: "Yes as. Hairy...very stubby neck, short neck, thick neck...feature the nose, nose kind of flat, and head is forward...kind something between a monkey and a bear." (He used his hand to indicate forehead.)

Steenburg: "Did you notice anything...is there anything you want to tell me about? Oh let's say, the eyes, the mouth...what was your impression about them?"

A: "Eyes. I could not see correctly, very clearly because the hair. Hanging down...the facial feature. It more resembled a orangutan...face."

Steenburg: "Could you describe the arms?"

A: "Arms? Very powerful shoulders, very broad shoulders, thick muscle, long arms."

Steenburg: "Are there any other characteristics of this creature that stood out in your mind?"

A: "The intelligence. The intelligence, because of picking berries, and doesn't destroy the berry growth itself. It's picking right from the top."

Steenburg: "So you're saying it was picking berries in a very humanlike manner?"

A: "Humanlike manner."

Perez: "Can you describe specifically, with its hand, how it was picking those berries?"

A: "Picking like most human beings, except a little bit awkward. Because it pulled the branch, but don't touch the berry, but brought the branch to the mouth."

Author's note: This is the same method of feeding that Mike McDonald described when he told of the sasquatch he observed eating leaves from trees in chapter one. It pulled the branch down and stripped the leaves with its teeth.

Perez: "So you're saying it actually wasn't picking the berry off the branch. It was bring the branch to its mouth?"

A: "Branch to mouth, yeah."

Perez: "And using the mouth to eat the berries?"

A: "That's correct."

Steenburg: "Could you tell if it was male or female?"

A: "I'm sorry, can't see that."

Steenburg: "How long did you see this creature for?"

A: "About...three minutes."

Perez: "Getting back to the gender momentarily, did you notice if the subject had breasts?"

A: "No. It was side, like in position where I could not see the front completely. It was backward, sideways."

Steenburg: "Did it ever make any noise?"

A: "No. No noise."

Steenburg: "Did it see you?"

A: "Yeah, in eye and ears."

Author's note: What Joseph meant here was that it both saw and heard him.

Steenburg: "What was its reaction when it saw you?"

A: "The reaction was it look at me right in the eyes, turned and looked to the left and look at me. And not nervous reaction, but start going backwards. Kind of surprise minute, and after a couple of feet, it started speeding up and go faster."

Author's note: I believe what Joseph was trying to say here was the creature turned its head to look at him, and while looking at him it backed up a few feet before turning and walking off. Again Joseph's thick accent made him hard to understand at times. Also, earlier in the interview I asked him if he could describe the eyes; he said he could not see them clearly due to long hair and the fact that the creature was not facing him. It now seems the creature did look at him, when it realized it was not alone. I suppose the hanging hair could still be an explanation here.

Steenburg: "So it basically left the area?"

A: "Left the area."

Steenburg: "Did you smell anything?"

A: "Yes. Kind of...foully odor."

Steenburg: "Just foul?"

A: "Foul, like a sulphur, sulphurish odor. Something like rotten something. Something like that."

53

Perez: "Where have you smelled sulphur before?"

A: "There's plenty of sulphur in this area. The hot springs here, there's hot springs out there too."

Steenburg: "After the creature moved off, did you see any footprints?"

A: "Yes, but there was leaves, dry ground and lots of leaves, you could see footprints temporary there, but then went back there to investigate further...then I tell Mr. Dahinden to show him the footprints was because of the leaves kind of covered the footprints. On the ground there was too many leaves and the leaves don't leave much."

Perez: "How did you know how to get hold of Rene Dahinden?"

A: "He gave me his address. His address in...Richmond."

Steenburg: "Did Mr. Dahinden come out and investigate?"

A: "Yeah, he come out. By coincidence, actually, he come out first."

Author's note: When Joseph first came to my trailer he stated that he had never reported his sightings to anybody. Here he said my colleague Rene Dahinden came to look at some foot impressions in the leaves. A bit of a contradiction here. When I asked Rene about it, he did recall it vaguely. Rene was in the area at the time this sighting took place. John Green and the late Robert Titmus, residents of Harrison Hot Springs, were not informed of this sighting at the time.

Steenburg: "Did you report what you saw to the police?"

A: "No, I didn't. Impossible to report from there...isolated area. It's one-hour drive from here."

Author's note: He means one hour from the town of Harrison Hot Springs. It is interesting that he felt he had to justify not reporting to the police.

Steenburg: "Okay, with the exception of Rene Dahinden, did you report it to anyone else?"

A: "I talked to co-workers about what I saw, but there were all skeptic."

Steenburg: "My final question. In your own words, tell me the story again from beginning to end—what you were doing and what happened."

A: "This time I am camping, by the Silver River and Harrison Lake mouth. And that season, looking for berries to make wine. That now I...looking through the patch, the big patch there. There are berries and I start picking berries there, several times. That sightings happened, when the berries were ripe all ready. Very well, so there was plenty of them. Like before, it was berry here, berry there. But this time it was plentiful. Every second day the berries was ripe, every second day! Collect more berries or eat them or anything."

Author's note: Joseph was essentially saying it was a bumper crop for berries that year. He continues.

"And that creature one afternoon between 4 and 5 o'clock, around there, and I noticed that something was different, that is something different as a bear. Because so gentle. So gentle with the bushes, with the berries, with the berry bushes, how it pick it up. Bear doesn't pick it up that way. Bear put everything on ground and eat it from ground. And that creature, take a bush (branch) with the berries on it, and eat only the berries. And not underripe berries. So there is some intelligence, so I discarded it as a bear, a bear eats everything. I was kind of shocked cause I could not describe (comprehend) what it is. I was not sure, was it a bear or something, something different. So I discarded (dismissed) as a grizzly bear, cause a grizzly bear would be acting more...aggressively. And that creature was gentle. Also grizzly bear would make certain noise, and that doesn't. Grizzly bear usually...getting...bark, like a dog, and that doesn't. I thought in my mind, I'm seeing something that many people want to see it, and I want to tell everybody then, but very few people will believe what I'm seeing. Because that time, lonely up there, maybe I'm just seeing things. After that incident I stayed up there another five years, I applied with the logging company to get a job, so I further investigate that situation. And after no sighting for a few years, I find footprints in the spring time in the melting snow, which was like human footprints, except way larger and way wider."

Author's note: This was the first time during the interview that Joseph mentioned finding footprints in snow two years later in the same area. Needless to say, I now had more questions for him.

Steenburg: "How many years after the sighting did you find these tracks?"

A: "About second years (two), that was on the road to natural hot springs."

Steenburg: "Did you photograph them?"

A: "No, I didn't. No camera, cause I was not prepared to see something like that. And I thought that maybe some crazy man from city come up, and walk bare foot in the snow here. That's what I thought. And that was not the case, because when I measured the footprints...that was BIG! And there was more like uh...resembled a human foot, except the heels, uh...so wide, almost like front of foot. So the larger heels, larger length, larger length, so that doesn't add up with the human footprint. And nobody would walk, anybody with sound mind, in snow barefoot. So I discarded that. I went further up and the footprint from the top of the...northeast from Harrison Lake, which is toward the hot springs, and then to the mountains, rocky area, which several caves there, several caves there. And I suspect that creature stays somewhere there in the caves—in winter time."

Perez: "Did you ever go to these caves?"

A: "Yeh, I investigate them, there were leaves inside. Which is unusual because I hardly believe the wind would blow the leaves so deeply in the caves, eh."

Perez: "Did the leaves look like...an area where the thing was bedding down?"

A: "Yeh. They were flattened. They were flattened."

Perez: "Did you get the impression that something was sleeping or living in there?"

A: "Yes, because there was no spider webs around. There would be spider webs, doesn't matter, early in the spring, or late in the summer, the spider web is always there! Right in the mouth of the cave."

Perez: "Did you observe any footprints in the cave?"

A: "No, you couldn't. There was...it's hard rock and leaves, you could not see the footprints."

Perez: "Did you observe any feces or urine in the cave, or smell urine?"

A: "Yeh. Feces yes. Outside the cave. Outside, at the mouth of the cave."

Perez: "Were they easily recognizable feces?"

A: "Yeah, it look like bear feces, but not quite."

Perez: "Did you suspect maybe a bear was sleeping in the cave?"

A: "Yeah, I suspected that first, yeh. But later because I know how bear feces...but bear is still sleeping at that time and it was early spring. That was in March yet, the snow just started melting."

Perez: "Have you ever smelled urine from a bear?"

A: "Yes."

Perez: "And the urine you smelled in the cave, did it smell like a bear or human?"

A: "No. Way different, it's way different, it's rotten egg smell a kind. Way different smell."

Steenburg: "Back to your sighting. Did you see it first or smell it first?"

A: "No, I see it first."

Steenburg: "When did you smell it? After it had gone?"

A: "After it started moving, then I smell it. I presume it had some kind of glands for self protection, which in stress...like many animals have scenting glands in the ears. Or behind legs or ah...some other animals have them in front."

Perez: "How close were you to the subject?"

A: "100, 150 feet, not far."

Perez: "Did you sense that you were in danger?"

A: "Uh. It was more hypnotizing than anything else. It like earthquake hit you. You don't know what to do in certain situations, because somehow interfere with you, function of your nervous center, nervous system in your mind. That creature have effect on my nervous system, on my mind reaction."

Author's note: Joseph here was simply stating that the creature left him shaken and confused. He was not trying to say the creature had some sort of mind control over him.

Perez: "Did you think these things we call bigfoot or sasquatch existed before you had your sighting?"

A: "Yes I believed it. But I was always skeptic till I saw that. Then I started more believing, and after that, five years. I spend the time to study it more."

Well, as you can see this was quite a report. Whether or not it is true remains the biggest question of all. Of course the same goes for all reported sightings of the sasquatch. Joseph did give us a full interview on his second sighting, which I will go into later. The next day the three of us went to the location of his first sighting and to the cave he claimed to have seen footprints near. It took us about an hour and a half to drive there, up the forestry road that runs along the east shore of Harrison Lake, up to the mouth of the Silver River.

We found a logging camp set up very close to where this incident took place. Joseph told me that the camp was shut down and unoccupied in July of 1988. I do know that this is common practice with the B.C. logging industry. When a camp is not in use, they will have one man living there to watch over things until it is needed again. Or in the case of a camp not too far from some community, somebody will be hired to come out once a week or so to make sure everything is all right.

The cave where he said he saw footprints was only about nine feet from the roadside. I used Daniel Perez's spotlight and went a little ways inside. I discovered that this was not a cave at all, but a long-abandoned mine entrance left to the elements. Thick brush had by this time almost covered the entrance so it could be easily missed by passing cars. The remains of old wooden rails were still there on the floor. After having a good look around we proceeded into the bush on the other side of the road to the mouth of the Silver River, and to the spot where the 1988 sighting took place. We found Joseph was about only ninety feet from where the creature was feeding.

The whole area was covered in large trees and the berry bushes when covered in leaves would make a very thick wall of vegetation indeed. We were there in early May, so it wasn't so bad. It would also appear that Joseph was slightly uphill from the creature when he spotted it. This could in my opinion have made it appear shorter than it actually was. Joseph said it was more than six feet tall. I think it could have been taller than that, perhaps seven or eight feet tall.

However, this is speculation on my part; maybe the creature was just over six feet tall.

Joseph also gave us a detailed interview concerning the second sighting he had on March 30, 1993. Again he was not carrying a camera with him, a bad habit I hope I've convinced him to stop doing from now on.

Perez: "Can you tell us about your second sighting, which happened in 1993?"

A: "Yes. That was on Hornet Creek Road."

Perez: "In British Columbia?"

A: "Yeah right, right, the lake, same spot we're talking about."

Perez: "How far from the first sighting?"

A: "It's only about more or less five miles."

Perez: "Okay, go ahead and continue."

A: "Five miles. And the first I know...because that time of spring time, I was helping my friends to cut...shade blocks."

Author's note: Just a translation here. He was cutting shake blocks. He was looking for big cedar trees, fallen down but not rotten, that could be cut in short sections from which shakes (long thick, untapered shingles) could be split."

Steenburg: "Before you go on. For the second sighting, just tell us what date did the second sighting take place?"

A: "March. About March 30th."

Steenburg: "And the year?"

A: "In '93."

Steenburg: "Was this one at night or day?"

A: "Daytime."

Steenburg: "What time of day was it?"

A: "Ah...early afternoon, about, between 1 and 2 o'clock."

Steenburg: "Okay, now describe the area in which this one took place?"

A: "That one took place, cedar growth and, and ah...clear logging, eh...coincide, like clear logging here, and cedar growth, ancient cedar growth just beside it, and ah...a clear-cutting area...what do you call it? Devil's claw bushes growing. It's abundant there. It's kind of moist area. I noticed then that several devil's claw bushes, which is very horny, you cannot pick it up with bare hand, was torn out of the ground. Who the hell would be of mind to dig the devil's claw? Because that's a dreaded kind of plant to everybody there, cause it goes in your flesh and you cannot get it out. That's why they call it devil's claw. I noticed that the roots were consumed. The roots were taken out and the bark. The bark was taken off the roots, it looked like a bear chewed on it. Something like that, and ah, the rest of the plant was discarded. And I went further, the devil's claw, the first one in spring time who start...getting greens on the top. Devil's claw, humans use it for rheumatoid arthritis, as a medicine. Extremely valuable medicine for humans. Indians knew that for thousands of years."

Author's note: The plant he's referring to is actually called devil's club. It is a thorny bush, fairly common in the West Coast mountains of British Columbia, as well as Washington and Oregon.

Steenburg: "What distance would you estimate you were from the creature when you saw it?"

A: "Two or three hundred feet."

Steenburg: "What was your first reaction this time?"

A: "Same reaction, I thought it was a bear again."

Steenburg: "What was it doing?"

A: "Picking at the devil's claw bushes."

Steenburg: "Did it stand and walk on two legs?"

A: "When I first saw it, it was four-legged, because eating, chewing the bark off the roots. It had it in the mouth and was holding it with two hands and eating it like this (he demonstrates, like corn on the cob). More like a monkey, something between a monkey and a human. Bear doesn't do it. Only two creatures do this; one is raccoon can do this and the other one is squirrel. Eat like this (he demonstrates again)."

Steenburg: "Was it covered with hair?"

A: "Yes sir."

Steenburg: "What color was it?"

A: "Something, brownish, reddish color."

Steenburg: "How tall do you estimate this thing to have been?"

A: "When it stand up after eating roots, and go to get other root out, other plant out of the ground. It was...way over six feet...tall."

Steenburg: "How heavy do you think it was?"

A: "300 or 400 pounds, around there."

Steenburg: "Did you see any facial features this time?"

A: "No. This time again because I was back (behind)."

Steenburg: "Is there anything about its physical characteristics that you would like to tell me?"

A: "Extremely big muscles, extremely powerful creature."

Steenburg: "Could you tell if it was male or female?"

A: "No, I couldn't. From that distance, no I couldn't."

Steenburg: "So it didn't have female breasts?"

A: "Female characteristics, no."

Steenburg: "How long did you see it for?"

A: "I observed, about four or five minutes. At least until it noticed me, because I fell, because the snow was just melting, and I fell in a hole there and start screaming, ha, ha, and the tree trunk I went over."

Steenburg: "Did it ever make any noise?"

A: "No. Not that time, no noise, no."

Steenburg: "Did it see you?"

A: "Yes. Certain because it turned to where noise was coming from. That's where I see, temporary, a second feature."

Steenburg: "When it saw you, what was its reaction?"

A: "Reaction? First start walking four legged, then after two legged."

Steenburg: "Did you smell anything?"

A: "Yes, you could smell again. Same type of odor."

Steenburg: "Same?"

A: "Something odor. A little different than the first time because maybe the animal was sweating that much? There was not that bad, sulphur in it. Not like that rotten egg. Probably not sweating, it was still cold yet."

Steenburg: "After it moved off, did you check for footprints?"

A: "Yes I did."

Steenburg: "Did you find any?"

A: "Yes sir, went by in the snow patches."

Steenburg: "What did they look like?"

A: "They look like a human, human footprint. Except it's wider and way, way longer, and have toes too."

Steenburg: "Did you have a camera with you this time?"

A: "No, no camera, because working on shaves there, cedar."

Steenburg: "When you smelled it this time, did you smell it before you saw it, during or after, or all three?"

A: "After, after when it take off. I smell."

Steenburg: "Did you report this incident?"

A: "Yeh. I report it to people working not far from there."

Perez: "Did you also notify Rene Dahinden?"

A: "No, I did not. No."

Perez: "Why?"

A: "Because somebody borrowed the book, he gave me the address, he gave me the phone number, and somebody borrowed for reading and never returned it (Sasquatch, by Hunter & Dahinden)."

Steenburg: "In your own words, describe the second incident for me."

A: "That creature. Amazingly cautious creature. Resembles a lot a bear, but it's not bear. It's something between ape and human. Because the intelligence again is there. Able to pick up certain roots of medicine value. Again careful with everything—everything around them, everything observed."

Author's note: I think here he was trying to express how impressed he was with how alert he felt the creature was. He continues.

"Walking off long, long steps, like two legged. When they walking distance, walking two legged. Only I see taking one step here, then walking four legged."

Author's note: When Joseph describes four-legged movement, I think he means it was moving about on its knuckles as it was feeding on the devil's club roots. He continues.

"Till it get to steep rocky area it was walking two legged. Quite fast! And it was a little bit different from the first sighting, cause that creature has a little bit darker brown color than the other one, the first one. Kind of more...aggressive looking. More aggressive looking. More aggressive behavior. Sitting up, more panicky, as the first creature was. Probably hungry, and also it was the mountain, and the foots was in the snow patches. Lots of snow patches that time of year. It's amazing something...it is hypnotizing and scary all of humans."

Author's note: I think he's describing how he felt while watching the animal, but I'm not sure. He continues.

"For me it was about 100 yards. That's what his creature all about, because they notice any other animal, and they bark and run after them, but not after that one."

Author's note: It now almost sounds like he's describing how dogs react, where this comes from at this point in the interview I don't know due to the fact he did not mention having dogs with him before. He continues.

"When I come back to other people, the caravan there, there was Andy, and...Larry! They said my face was white. What happened to me, I was scared. Not only because I fall in a hole there, but I get scared at what I saw. I noticed the dog was lying down by the car, lying down flat! Scared! The dog is scared!"

Author's note: Now he mentions dogs, my confusion over his earlier statement now goes away. He continues.

"Noise and the dog is scared."

Author's note: Earlier when I asked if it made any noise he told me no. A little bit of a discrepancy here.

Steenburg: "Did any of these people see it?"

A: "No, they didn't, they see something because they hear that, and they went up the hill, the rocks was clamping? They hear something was happening up there."

Author's note: Joseph, I think, is saying the people below heard what might have been the sounds of rocks banging together. This behavior of sasquatch banging rocks together has been reported before.

Steenburg: "Did they see footprints?"

A: "Yes they did."

Perez: "Did they believe you?"

A: "Yes they did first, yeah they did believe me. But they say no sasquatch, it must be bear. They don't believe in sasquatch, but I say for sure was sasquatch!"

Steenburg: "Did they say it was still a bear even after they saw the tracks?"

A: "Yeh. They say grizzly bear has big tracks, ha, ha."

Perez: "You said your March, 1993, it was more aggressive?"

63

A: "More aggressive."

Perez: "Please explain."

A: "Moving faster and more panicky. Kind of mad (angry). Throwing rocks backward, rocks like with hand."

Author's note: This was the first time he mentioned seeing the creature throwing rocks.

Perez: "You also said about the March, 1993, sighting that the subject was extremely cautious. Please explain."

A: "Yeah, cautious like, anything to do, it first looked around. What was around it. Soon it do certain chores. What it was doing, eating or picking up roots? It did certain things in split seconds, then look around and look around. Then eating, still looking steadily around, the head moving back and forth, right and left. And usually the eating...on the ground...that time...eat, what the creature was, it's eat on the ground and hold it. The...devil's claw root, with the paws, or hands like? (Like a person with a cob of corn.)"

Author's note: Again like Mike McDonald in chapter one, the creature seems to be always scanning its surroundings as it feeds— a very alert and high-strung creature.

Perez: "You said the subject you saw was extremely well built and muscular?"

A: "Very muscular."

Perez: "Underneath the hair, how did you notice the muscularity of it?"

A: "The thickness of the legs and the forearms. (He says something else here that I can't make out. His accent again causing difficulties.)"

Perez: "Did you see muscle definition or just a bunch of mass?"

A: "Mass, it's mass. You can see on the legs there the definition of mass because of the legs...in the back of the legs, less hair."

Perez: "Was it built like someone who lifts weights or was it a natural build?"

A: "Looks like somebody doing a heavy lifting job like."

Perez: "Not like a body builder?"

A: "No, no."

Perez: "Just built heavily?"

A: "Yeah, around there, yes."

Perez: "Do you think the subject you saw, March, 1993, could have been the same one you saw earlier in 1988?"

A: "Could be. But I'm not sure because the color of the hair was different. The first one in 1988, the color of hair was lighter. This one was darker, brownish color. I don't know if something...temperature could have something to do with that, or just because the nutrition changes? Because a lot of animals, nutrition has to do with the shine of fur, and does the temperature have something to do with an animal have a darker color, in the winter to absorb more energy from the sun?"

Perez: "Did you think the characteristic of the subject that you were looking at were more apelike or manlike?"

A: "Something further from ape...ape and man. Close to Neanderthal, like you see on television, something close to that."

Steenburg: "In your opinion, do you think this was the same creature you saw earlier or a different one?"

A: "It was a different creature. It was more aggressive, color different, and also that one walked four legged when it reached the mountain side. The steep sides."

Steenburg: "It was using its hands to help itself up?"

A: "Help itself up, yeh. The other one I don't see it use hands."

Steenburg: "The first one (1988), did it go up hill when it went away from you?"

A: "Went uphill. But not as steep as this one! That's the difference."

Perez: "The second hill, in terms of degrees (Daniel uses his arm), if this is 90 degrees, what degrees?"

A: "It was about 45 to 60 degrees, you see, aound there. Rocky, very rocky."

Steenburg: "Rocky in both incidents?"

A: "The first one not that rocky, no rocky, lots of leaves, that was close to river. This one was up the hillside, the mountain."

Steenburg: "The first one, what angle was it at when it started to go uphill?"

A: "That was only a little bit till it hit the road, then it disappear."

We did try to go to the location of Joseph's second sighting, but about halfway up the Hornet Creek Road, we found it blocked by a rock slide, as well as being washed out in some places. Being early

May I assume work crews just hadn't been out yet to repair it. Logging companies usually don't repair roads until they are needed again. Since our interview with him, Joseph has appeared on some television shows about the sasquatch, one of which I now have a copy. In this show they reenacted his 1988 sighting with an actor in a costume. However, the show made no mention of his second sighting in 1993.

If he was telling the truth and did indeed see sasquatch on two occasions five years apart, I hope like hell he took my advice to heart and now carries a camera with him whenever he leaves Harrison Hot Springs.

Right: Joseph Verhovany stands at entrance to cave, where he is reported to have found footprints. On inspection, the cave turned out to be a long-abandoned mine entrance.

Below: Joseph with Daniel Perez inspecting the area at the location of his 1988 sighting.

Photos: T. Steenburg.

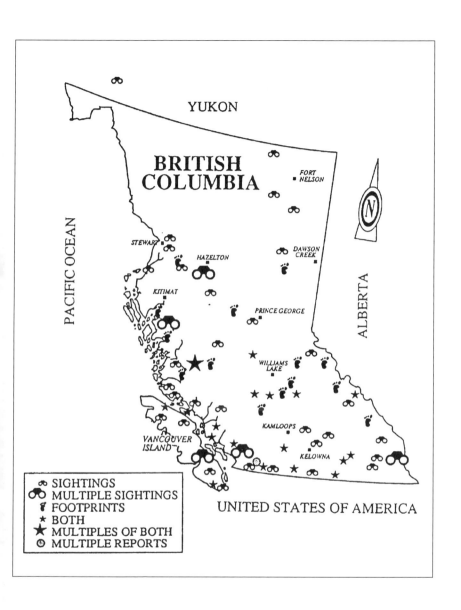

YUKON

BRITISH COLUMBIA

FORT NELSON

PACIFIC OCEAN

STEWART

HAZELTON

KITIMAT

DAWSON CREEK

PRINCE GEORGE

ALBERTA

WILLIAMS LAKE

KAMLOOPS

VANCOUVER ISLAND

KELOWNA

🐾 SIGHTINGS
🐾🐾 MULTIPLE SIGHTINGS
👣 FOOTPRINTS
★ BOTH
★ MULTIPLES OF BOTH
◉ MULTIPLE REPORTS

UNITED STATES OF AMERICA

Two examples of how the sasquatch is a key part of Harrison Hot Springs culture. I don't think any of the towns residents would try to erase the sasquatch legend from their community today.

Photos: T. Steenburg, 1993.

More Encounters with B.C.'s Giants

Although I wrote a good deal about the Harrison Lake area in the last chapter, I do possess a few more reports from other people in that area. One of which was given to me by a Calgary friend who grew up in the Harrison Hot Springs-Agassiz area. Rick Doc (not his real name) lived with his mother, father and brother in a house near the Harrison Hot Springs elementary school in the late 1950s and early 1960s. He now works at a hospital in Calgary and has been a good friend for many years now.

Knowing my interests, he would frequently ask questions about what was happening lately in the sasquatch field, and he always seemed to have a keen interest in my research. I often asked if he knew anybody or had a sighting himself, growing up where he did. He would just shake his head, and say no—too quickly I thought at times, but I never pressed the matter. His wife Sue, a wonderful woman, knew her mate was holding something back I suspect. But she never said anything, although I felt she wanted to many times when I was visiting their home and the subject came up.

I was at their house one evening in 1995 when, as usual, the subject of the sasquatch came up. After some discussion Rick and Sue grew silent. I watched them look at each other and then Rick said to me, "Buddy, I've not been honest with you about seeing a sasquatch myself and I'm sorry about that." I told him he had nothing to be sorry for and if he now wanted to tell about it, I was all ears. He then went on to tell me that when he was a boy, he and his brother saw what they thought was a sasquatch in the hills behind the Harrison Hot Springs Hotel, during the summer of 1958.

The two boys were out having fun exploring the woods. Suddenly, they both heard a rustling noise ahead of them. The two boys froze and just stared ahead trying to see what was making the noise. It was then they both saw a large, hair-covered creature just behind the trunk of a large old-growth cedar tree about 100 feet ahead. The creature was mostly obscured from view behind the tree, so they only saw the back half of the creature's right side, in profile. The front of the creature was behind the tree. Its right arm and hand though could be seen clearly, as the creature seemed to be holding onto the tree. Rick did not know if the creature was trying to hide itself from them or if it was observing something downhill ahead of it toward the springs. Rick's brother (Gary) at this point panicked and took off back in the direction from which they had come. The creature at this point also took off; all Rick could hear was the thing running away from him as it moved down the steep hillside. He couldn't see the thing running, but he heard it. When Rick got back to the point where he and Gary had first walked into the woods he found his brother there crying. He was worried about what had happened to Rick. He also probably felt bad for leaving him there alone with this strange creature. Later that day Rick told his parents about what they had seen, but he was not taken seriously.

The next time I visited Rick and Sue at their home I had my tape recorder and my standard questionnaire with me to do a proper interview.

Q: "Would you state your full name please?"
Real name withheld.
Q: "Where did this incident occur?"
A: "Harrison Hot Springs."
Q: "What date did this take place?"
A: "'58, '57, I think 1958."
Q: "What time of year was it?"
A: "Summer."
Q: "Do you remember the month?"
A: "July."
Q: "Was it at night or day?"
A: "Day."
Q: "What time of day was it?"
A: "It would have to be 4 or 5 o'clock."
Q: "Describe the area in which this incident occurred."
A: "Deep foliage, heavy trees, big trees."
Q: "Where about in Harrison did this happen?"
A: "Near the hotel, above the hot springs."
Q: "Closer to the hot springs or closer to the hotel?"

A: "Closer to the hot springs."

Q: "In the hills behind there?"

A: "Yep."

Q: "What distance would you estimate you were from this thing when you saw it?"

A: "From here to my fence (pointing to a backyard fence)."

Q: "About 100 feet?"

A: "Yeah, I think so."

Q: "What was your first reaction?"

A: "My first reaction was, 'Holy shit, what's that?'"

Q: "What was it doing?"

A: "Standing upright against a tree."

Q: "Were you alone when you saw this?"

A: "No, I was with my brother."

Q: "What's his name?"

Real name withheld.

Q: "Did it stand and walk on two legs?"

A: "It stood."

Q: "Did you see it go down on all fours?"

A: "No."

Q: "Was it covered in hair?"

A: "Yeah."

Q: "What color was it?"

A: "Brown, grayish brown."

Q: "How tall would you estimate this thing to have been?"

A: "A little higher than my door (pointing to his backdoor frame), which is..."

Q: "About seven feet?"

A: "Yeah."

Q: "What would you estimate its weight to have been?"

A: "About the size of a grizzly."

Q: "So heavy?"

A: "Big, yeah. It had elbows."

Author's note: At this point Rick started remembering details. It was almost a flood of memories coming back to him now and he would give details before I asked him. He continues.

"It had elbows when I think about it now. It had elbows, because a bear does not have...well, I guess...it looked like...elbows...you know what I mean? When it's bent, a bear has an elbow, but it has a longer span between the elbow, than a bear. A bear is kind of...I know what a bear looks like, and it wasn't that, because you can tell. The joint from the shoulder to the elbow was longer, you know? And stringy hair, not tight hair like a bear would have."

Q: "Did you see any facial features?"

A: "No."

Q: "Did it have its back to you?"

A: "Half and half. Its head was turned, but its back...and this part of it (runs his hand down his right torso to demonstrate). Its head was on the other side of the tree itself. It was like it was looking for something there.

Q: "Could you describe its arms? You were talking about the elbow?"

A: "The elbow, like I say, its from the shoulder. Now if you look at it...look at a bear right, standing up and sniffing something, you will notice that the bear has a big hunch (hump) where the shoulder is, and you can see the movement there. But you also see between the elbow and that it's wider coming out this way from the paw or from the elbow to the hand...or the paw...is wider. That is the difference that I saw. But this thing I saw, it was straight arm and longer. And yet the elbow to the shoulder was definitely not a bear's. It wasn't bent like a bear. There's no way...I mean it was formed round, where a bear seems to have a point. That's what I saw. To me it did not look like a bear. And it stunk, real bad, and a bear stinks, but not like that."

Q: "Could you see if it was male or female?"

A: "No."

Q: "Did it ever make any noise?"

A: "No. From that distance I just heard rustling. There were bushes right where it was standing, and it was moving that stuff around."

Q: "The animal itself never made any noise?"

A: "I didn't hear it, no. From that distance, no."

Q: "Did it see you?"

72

A: "Well...I think it did after I yelled, because that's when it started to take off. But when it took off it went...the angle it was on, it went behind another set of trees. I went around like this (demonstrates with hands) to take another look around the side of the tree, from that distance I had, and it was gone, almost, like it was gone (hands up, shrugging his shoulders). But when it took off it was on two legs. It was not crawling."

Q: "When it took off, was it heading down towards the lake or up on the hills?"

A: "Going down towards the lake. It did not look like it was running. It looked like it was trotting."

Q: "Did you smell any odor during this sighting?"

A: "Oh yeah, pretty strong. At first I thought it was a bear. Bears, they smell? It smelled...putrid, like somebody upchucked or something. It was pretty smelly."

Q: "After it had moved off, did you check for footprints?"

A: "No. We were too scared."

Q: "Did you report what you saw to the police?"

A: "No."

Q: "Did you report this to anybody?"

A: "Just to my parents."

Q: "What did they say?"

A: "Oh yeah, right!"

Q: "They didn't believe you?"

A: "No."

Q: "In your own words, tell me what happened."

A: "Well...me and my brother decided to go looking around the caves—stuff like that. We thought we would go up by the hotel, check out the hot springs, dip our feet, stuff like that. Anyway, we went a little further into the bush. There was this trail, a little bushy, but hey let's follow it. When we got to a certain point there, we heard this rustling, and we stopped. We looked around. Didn't hear anything behind us. We looked to our front and that's where we saw this thing, standing there, literally standing there. Another point here, the legs were not bent like a bear because when a bear stands, he's bent, and this thing was straight legged. And the back end of it? There was no tail. It was just flat, straight flat back. That's when I said to my brother, 'What the hell is that?' And he said, 'I don't know but let's get the hell out of here!' I said, 'No, let's get another look.' Well, my brother took off. I went a few more feet, but when I went behind those trees it seemed to disappear in...it was gone, and there's a gully, sort of like a drop off down there where I think it went. It must have rolled or whatever. I thought to myself, this is pretty damned hairy (scary). So I went back and got my brother, he was crying, because he was worried about me. I was supposed to watch over him right.

"When I got back he actually didn't want me to tell mom and dad, but I told them anyways. He (brother) had bad nightmares that night, but he wouldn't admit it right. My brother's like that. We went back the next day. We thought we would go down to the gully and work our way back up to the same spot. I didn't see any footprints. Of course, at that age we weren't looking for any footprints, we weren't thinking about that. We were just looking to see if we could see it. We did see on the tree where it was these scratches. Now these scratches weren't deep like a bear's. Now these scratches were...flat, where a bear digs deep, you can see the deepness of it. These were flat—a light scratch. I bet to this day if I could find that tree, if it's still there, if they haven't cut it down. I know where that spot is, when I think about it now, because you're bringing back memories again. So I'm sure I could find it again. If it's not there, it's because either the tree has been taken away or the bark has dropped off or whatever. But I know the area. If we could go there, I would show you.

"So that's what happened. I never went back up there. I never ventured any further, me and my brother never did, we just stayed in town. I think because it bothered us so much. It wasn't a bear! That's the funny thing about it. I know a bear would fall down, look at you, smell you, then leave. This thing just stood, then walked away around the tree, and just took off, like it was gone before I knew it! At least that's what I remember."

After the interview Rick seemed relieved about having finally told me about this incident from his boyhood. He then drew me a sketch of what he and his brother saw that day in 1958.

Was it a sasquatch or was it indeed a bear? He is wrong in some of his statements of bear behavior. Bears do indeed have the ability to stand straight up. And since he never really watched the creature as it moved away, how could he know that it wasn't moving away on four legs? One thing does make me believe it was a sasquatch he saw that day. He was adamant the creature had a hand on the end of its arm, not a paw.

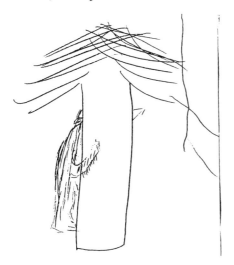

Sketch drawn by Mr. Rick Doc (not his real name) of creature he and his brother saw in the hills behind the Harrison Hot Springs Hotel during the summer of 1958. The creature was partly obscured behind an old-growth cedar tree, with one arm in plain view. Rick was adamant that the creature had a hand on the end of its limb, not a paw.

Although most reports of encounters between humans and sasquatch that come my way occur on the B.C. mainland, I've noticed over the years now how few I've received from Vancouver Island. Why this is, I have no idea. I do have turn-of-the-century reports, as well as a number from the 1930s, but very few in the last forty years or so. It could very well be that I've just not heard about them, for I know other researchers have had quite a few. One of the few reports that has come my way, came in a letter from Manitoba researcher Guy Phillips. Guy was quite active in the sasquatch field during the late 1980s, and early 1990s. However, it's been about four years since I've heard from him. I don't know if he's gone on to other things, and stopped looking into reports or if he's still active. Hopefully he will read this book and make contact again to let me know what he's been up to.

Guy did send me a letter in 1988 concerning a fellow named Julius Szego, who claimed to have had a sighting of a sasquatch at a place called Scott's Falls on Vancouver Island, located along Route 18, north of the city of Victoria, between Duncan and Lake Cowichan, Guy's letter was as follows:

Found another B.C. item that may be of interest to you. I don't think I mentioned it, as always you are welcome to it. It took place in 1975, in the summer, Vancouver Island at Scott's Falls, (could be Scutz or Skutz. I went there, but like an ass neglected to record it). A guy named Julius Szego and his cousin Ed, about 15 years old at the time, saw a tall creature, about 7' tall, crouched at the edge of the water pulling out roots. It was on the same side of the falls as they were. It was brownish, black in color and had mossy hair on the body. It had no hair on face and a flat forehead. It had a human like face and when the boys had watched it for about 30 seconds, they threw a rock near it to grab its attention. The creature stood up looked at them, grunted, and walked off. They were about 40 feet from it and it was near the edge of the bush. This was about 8 p.m. They ran down and alerted the R.C.M.P. and Rangers who came back and took a look around with little result. I spoke to Julius in 1980 and spoke to a couple of his buddies who believed him. When I was out there, people in the area said there was lots of wildlife around like lynx and puma, which was supposed to be long gone from that area of the Island.

A little later I contacted John Green, due to the fact I could not locate Scott's Falls on any of my maps. John wrote back to me telling me he, too, had trouble finding the location. We both knew that a waterfall would have to be pretty spectacular in order to appear on any road map. When I wrote back to Guy informing him of our difficulty in locating Scott's Falls, he replied with a photocopy of his map with the area in question highlighted.

Later, I wrote a letter to the RCMP in Duncan in order to find a record of this incident—I was not expecting a reply. After about four weeks I did get a short reply stating that since it was now thirteen years later, any incident report of this nature would have been disposed of. Also, no officer posted at the detachment could remember the incident. However, they all had a good laugh when asked.

I can see why witnesses are sometimes reluctant to report their sightings to people of authority. Despite this, it has been my experience that even though a police officer may not believe in the existence of sasquatch, the RCMP are very professional and will, in most cases, fill out an incident report. The same holds true for national park wardens as well as forestry department rangers.

The only other Vancouver Island report I will mention here was not a sighting of the creature itself but a possible footprint find. In January, 1996, I received a long distance phone call from a man named Christopher Crocker, of Ladysmith, B.C. He told me he was out at Rineheart Lake, not too far from Ladysmith. While walking, he came across what he described as a clear, sharp, lone footprint of a sasquatch. I asked him if he took any photos of the print. He replied yes and he would send me copies of the photos right away.

After I hung up the phone I decided to call John A. Bindernagel. He is a wildlife ecologist in Courtenay, B.C., who, since I first met him in 1993, has been looking into reports of sasquatch on Vancouver Island. John agreed to look into this track report and get back to me. A few days later John phoned back and informed me he was not impressed with the alleged footprint, and as far as he was concerned the photo showed no footprint characteristics at all. It didn't even have toes! It's always handy to have a fellow researcher close to a report of this kind who can get to the site in a very short period of time. Now, whenever I hear of something happening on Vancouver Island, I now have a contact there who will look into the matter quickly. I don't know how Mr. Crocker felt about me sending John to look at his photos, but at the time of writing I still have not received any copy of the photos he said he would send. I don't believe Mr. Crocker was trying to perpetrate a hoax here. I think he truly believed he found something interesting, but as far as I'm concerned John put this matter to rest.

Another possible track find occurred near the town of Cranbrook, B.C., during the summer of 1975. Again, I found out about this find over the phone from a man who saw the prints. He did not want me to reveal his identity, so I will give him an assumed name.

Mr. James Benson (not his real name) went on to tell about several strange tracks he found as a fourteen-year-old boy while on a fishing trip with his father. While his father was fishing and cursing the bad weather that had persisted over the previous three days, James decided to walk down the river a little ways. The water in the Joseph River was very high and fast due to heavy rains. As he walked, he came upon large, humanlike barefoot tracks. He did not count them but told me there were quite a few of them.

"They were about fifteen inches long!" he told me. When I asked him how deep in the ground they were imbedded, he replied, "About an inch."

"What was the weather like that day?" I asked.

"Overcast and dark. The rain had finally stopped after three days of uninterrupted downpour."

It looked as though the rain was going to start up again at anytime, that's why his father was in such a foul mood that day. He followed the tracks to the river where the track maker appeared to have paused by the water's edge and squatted, as there were two impressions side by side, pointing toward the river. James called to his father to come and see what he had found, but his father had no interest and told him to get his butt back in the car as it was time to leave! James was not carrying a camera with him so the tracks were not photographed. On the drive back to Cranbrook, James pleaded with his father to go back and take a look. His father, still in a bad mood said, "What the hell is so important about some animal tracks anyway, it's not like you found bigfoot tracks or something!" He then laughed. James didn't say another word after that.

If Harrison Hot Springs, B.C., is the front door to sasquatch country, then the town of Bella Coola, located half way up the British Columbia coast, would have to be the livingroom. There have been many reports from this area since the turn of the century. They would make a pretty interesting book on their own. The following rash of reports appeared in an article from the *Coast Mountain Courier* in November of 1989, which was forwarded to me by John Green. The incident occurred on November 11, 1989. It reads:

On Remembrance evening, sixteen-year-old Jimmy Nelson, his friend Glen Clellamin and his mother were sitting around the kitchen table talking. They noticed a terrible smell and thought it might be a dead dog. Glen went to get a drink of water and that's when he saw it.

"There's an animal out there!" Glen exclaimed. Thinking that it was a bear on the back porch, after the deer meat they had hanging, they jumped up and looked out. They saw a large, hairy creature with long arms, approximately seven to eight feet tall, with wide shoulders, running like a human. They got a good look at the creature as their back porch light was on and it illuminated the yard. They were frightened and couldn't quite believe what they saw. The following night the dogs started barking. The creature was back and in the yard next door. This time the boys followed as it headed down the valley toward a creek.

They heard a growl and then a high-pitched scream that made the hair rise on their necks! Curiosity kept them walking slowly toward it. When they were about thirty feet away, it stopped and looked at them. They yelled at it, and the creature started walking toward them. Then they ran!

In the morning they went looking for tracks and found three sets. One large, one medium and one small—apparently a family. Other spottings of the family—sometimes the larger one, sometimes the smaller or medium one—have been reported by several people. The stories consistently seem to indicate a search for food and end with the creature running away on two legs. Its manner appears to be more leery than afraid. Pictures have been submitted to the Courier of footprints. As they were concerned the footprints wouldn't show up clearly, the tracks were filled with flour to produce a clear picture. Who is this new shy family in the valley? No one knows. But the village waits for the next sighting.

I did not learn of these reports until late December, 1989, when John Green sent me the article. He did not go there himself because of bad driving conditions at the time. Since no other reports have come my way about this sasquatch family, I assume sightings stopped about the time the article appeared in the *Coast Mountain Courier*.

Another secondhand report came from my friend, the late Vladimir Markotic. The witness in this case observed the creature from an eastbound train, as it slowly climbed up a hill through Glacier National Park. Miss Frances Rand, of Calgary, Alberta, wrote out a report, describing the creature she saw. She writes:

I saw a Sasquatch on June 16th, 1982, at about 1:30 p.m., from a moving train, on my way home to Calgary from Vancouver. It was in Glacier National Park British Columbia, between Revelstoke and Golden. The Sasquatch was about 75 feet away at the edge of the dense forest sitting on a fallen log. I could see his left side, he* seemed to be looking at the back of the train, which, by the way was very slow at that particular spot. The Sasquatch was very tall and very lean. I would say 7 feet or more. I remember thinking that he was much taller than my son-in-law who is 6'3" and also very lean. He was all covered with short dark brown fur, that resembled

"Borg lining", used in winter wear. I noted the shape of his head, because I marvelled at the roundness of it, that fact eliminates a lot of wild game, I did not see any ears. He looked like a very tall human all covered with fur. His left hand was up to his face, as if he was eating something? His left leg bent at a 90 degree angle, his right one extended. I saw him for 4-5 seconds only. But what I saw was a Sasquatch, it could not have been anything else, of that I am firmly convinced.

(I refer to it as a he as I did not see any breasts.)
Frances Rand

I do not know if anybody else on this train saw the creature Frances described. She also drew a picture of the animal and sent it along with her report.

Sketch was drawn by Ms. Frances Rand; it is a creature she claimed to have observed from a moving train on June 16, 1982. The creature was thin in appearance, and appeared to be eating something as it watched the passing train.

I also have two detailed reports from the year 1974. I was very impressed with the witnesses in both these cases, not only for the detail in which they described their encounters, but also with their insistence about what they had seen.

The first incident occurred southwest of Mackenzie, B.C., during the last week of October. Barry Smith (not his real name) told me that while he was out hunting with his uncle he observed a creature, which when he first saw it he took to be a tree stump. Later when the creature moved, Barry took his rifle from his shoulder and fired a shot over the thing's head. He did not shoot directly at it because he still wasn't sure just what it was. As soon as he fired, the creature turned and ran down a steep hill at very high speed. Barry, a former track runner, was most impressed with the creature's speed and grace. The creature ran straight down, not at an angle as a man would do to avoid losing his balance. It ran upright on two legs the whole time it was in sight. Barry came to my home to be interviewed. When I met him, it was obvious to me that he had lived the life of an athlete. He stands 6'4" inches tall and was still in excellent shape. During the interview he answered all questions without hesitation.

Q: "Where did this incident occur?"

A: "East, southeast of Mackenzie B.C."

Q: "What date did this take place, to the best of your memory?"

A: "I had just come out of the Arctic, so...I think it was the last week in October, 1974."

Q: "Was it at night or day?"

A: "It was the middle of the day."

Q: "What time?"

A: "It would have been around 2 o'clock, bright sunshine, clear, very clear, beautiful day out. Very windy."

Q: "Describe the area in which this incident took place."

A: "We were up on Firetower Road, over looking the valley, we were up oh...I don't know...1,000 feet anyway."

Q: "What distance would you estimate you were from this creature when you saw it?"

A: "The first time?"

Q: "Yes."

A: "Well, when I first saw it, I thought it was a stump. I was about, oh...5 maybe 600 feet away."

Q: "What was your first reaction?"

A: "Well when I realized what it was, I was amazed. Especially when it moved."

Q: "What was it doing?"

A: "Just standing there, looking at me."

Q: "Did it stand and walk on two legs?"

A: "Well...it was standing upright. It looked like it was shifting from one foot to the other."

Q: "Did you ever see it go down on all fours?"

A: "Never."

Q: "Was it covered in hair?"

A: "Yes. Black hair."

Q: "How tall would you estimate this creature to have been?"

A: "Well, when he finally turned around and ran, I was about 200 feet away. So...I'm 6'4", and he was a lot taller than I was."

Q: "What would you estimate its weight to have been?"

A: "Oh. I can't tell you that, heavy."

Q: "Did you see any facial features?"

A: "No, I could see its eyes though."

Q: "Could you describe them?"

A: "Mostly white."

Q: "Could you describe the arms?"

A: "Big arms. When he turned and ran, his right arm...when he swung, he swung to the left to run. His right arm came around, and I bet it was bigger than my upper leg. His bicep was bigger than my leg...and I have big legs!"

Q: "Could you tell it was male or female?"

A: "No, I couldn't."

Q: "How long did you see it for?"

A: "Well, from the time I first noticed it and when I first realized that something was there, and up to the point where he turned and run...three minutes."

Q: "Did it ever make any noise?"

A: "No."

Q: "I assume it did see you?"

A: "Of course. It had to have been there quite a while watching me. The reason I first noticed it is when I brought down the rifle off my shoulder, I grabbed the sling and swung the rifle down in front of me, that's when it first moved. He moved in behind the tree."

Q: "Did you smell anything?"

A: "No."

Q: "After it ran away, did you check for footprints?"

A: "No."

Q: "Did you report what you saw to the police?"

A: "Yes I did. I reported it to the Mackenzie RCMP. I ran back to my uncle. He was hunting farther down the hill. I ran, I told him, and he laughed.... Wait a minute, we did go back to check for prints, but we didn't see anything, because when I ran up to where it was, it was gone, that's how fast it was moving."

Q: "What did the police say about all this?"

A: "The officer, well, he kind of chuckled to himself and said, 'Well we don't believe in them here.' Indifference."

Q: "Did you report this to anyone else?"

A: "No. Well, over the years at parties and stuff when people were talking about strange things, I would say, 'Well I saw a sasquatch in B.C. in '74,' and everybody would have a good laugh."

Q: "In your own words, describe what happened."

A: "Well, we had gone up to the firetower, which is quite high, and I remember it was a windy day. We drove back down to the first plateau, stopped. We were bear hunting. That's why I thought it might be a bear. I was walking along the ridge and my uncle Dean, he went further down the hill. I was walking north, northeast, kind of looking around, and at first I noticed something. Well, it looked like a big, black stump standing against another tree, and as I was walking along, it moved and light came between it and the tree. Again, I thought maybe it was just the wind moving the tree back and forth. As I got closer to it, I thought I saw it shift again. Well, I brought the rifle off my shoulder thinking maybe it's a bear, and that's when it moved back in behind the tree. That's when I realized that this was no bear. I started walking towards it, and it moved again, farther behind the tree. And that's when I brought my rifle up and took a shot over its head. This thing turned and ran, and it was moving very fast."

Q: "On two legs the whole time?"

A: "Yes the whole time. I was impressed because I was a runner and there is no way that I could run as fast as that thing did, and he was taking some pretty long strides, and as he was running it hit me that I was looking at a sasquatch."

Q: "What did you do after it was gone?"

A: "I ran up to where I saw him, and looked down over the other side of the ridge down the hill. I couldn't see it anywhere. One thing that surprised me was that there were no slide marks, so that meant that it just ran down that hill. There were no places where he skidded or fell or slid, like a man would running as fast as he could down a steep hill. I ran down towards my uncle and told him that I had just seen a sasquatch, and we went back up to where it was and never saw anything, so we got in the truck, and on the way back to town my uncle said, 'Let's stop in at the RCMP and report this.' Which we did, but I don't think they took me seriously, and I know my uncle didn't."

Q: "What's your uncle's name?"

Real name withheld.

Q: "Does he live in that area?"

A: "No. He lives in Moosejaw now."

As can be seen, this is a very detailed report. Barry wanted his name kept confidential so it would appear he's not seeking attention by making up a tale about seeing a sasquatch. One thing does bother me though, and it seems to have bothered Barry as well. I don't understand why there were no scuff marks at least on the hillside where the creature's feet touched the ground as it ran down a very steep hill. When I asked him if the hill side was tree covered, Barry told me there was the odd tree, but most of the steep hillside was dirt covered with small rocks. One would think there would have been some sign of the creature's passing. So I have to conclude that either Barry, for whatever reason, is making this whole story up or he indeed saw a sasquatch that day in late October, 1974. One last thought of mine though, if he had shot right at it, rather than over its head, this mystery of the sasquatch would be over.

At about the same time Barry Smith was shooting over a sasquatch's head near Mackenzie, B.C., another encounter was taking place almost inside the town limits of Princeton, B.C.

In 1986 I had placed an ad in the Calgary press, advertising my research as well as inviting witnesses to come forward if they truly believed they might have seen or found something of interest. Of course, there were the usual crank calls and hoax attempts. That comes with the territory in this field, I will talk about them in a later chapter. However, I was surprised at just how many people there were out there who really believed they may have seen something interesting, but had no idea how to contact anyone who would take the reports seriously.

Glen Boulier of Calgary contacted me to tell of an incident he had with his brother as the two went jogging early one morning on the outskirts of Princeton during the last week of October, 1974. This incident occurred at the north end of an old wooden bridge that spans the Similkameen River, at the town's northern edge. At the time of this incident, the town of Princeton was located at the south edge of the bridge. There was virtually no development at all on the north side in 1974. The same was true when I visited the spot myself for the first time in 1986. However, when I stopped there again in May, 1996, it seemed to me the town had exploded across the river as where there once were trees and forest there were small businesses and new houses galore. Also, the road across the bridge was being widened at this time. The old bridge was still there, but it looked like a sad, old relic from a slower past, which no longer fit the growing town's needs. It wouldn't surprise me to find it has been replaced by now with a wider, concrete bridge. I went to Glen Boulier's home to interview him about the creature he and his brother saw in 1974. Glen was glad to finally have an opportunity to tell someone about that day.

Q: "Tell me what happened that day."

A: "At approximately 8 o'clock in the morning we usually go jogging in that area and there's a roadway that goes around the airport. And at that time we had just gone around the airport. It's about three miles around and we were coming down towards the bridge. And the last 200 feet we decided to walk and just walk across the bridge and go to where we lived. And at that time like, we picked up on it by seeing something on the roadway, and I don't know its distance from us...would be about 150 feet from us. It wasn't too far. We had just seen him, he was crossing the road at the time and we didn't know what it was at the time, and it was fairly black or brownish black, and we saw him lift himself off the roadway and up onto...I don't know if it is a service road or whatever, but it only took him one step to get up it. He just lifted his whole weight up it, and on he went, and he started to walk up the roadway. Now, at that time I was really scared because he was huge, eh! I was really scared and by that time we had already...like we were at the bridge, we were just standing there and as far as we were concerned, it had not seen us yet. It had walked up the road a ways and my brother...I didn't know what he was going to do, and he just yelled at him and it turned its body around so you could see him, okay. Like he was looking at us, it watched us for only a few seconds and then it turned around and walked back up the roadway. It wasn't scared or nothing, it didn't act like we were any danger to him at all. But like I said, myself, I was really scared, and my brother, it didn't seem to bother him, eh.

"It took my brother about ten minutes to convince me to follow it up the hillway, and I didn't want to get involved with it because he turned around. Like I was saying, there was no doubt in our minds, because you would swear that you were looking at a gorilla or something like that. We decided we were going to go after him for a little ways, but he was far ahead of us, okay, and we were going from print to print in that skiff of snow. We could see its prints, so we were following the prints eh...and um, the print distance (step) I had to give a little hop eh, to get from print to print. His walking distance was anywhere from four and a half to five feet from step to step. Really large stride, and that was only walking, and the same when it went up the hillside there, okay, up the roadway.... I was trying to get his height or whatever, so I went up there and tried to touch a few of the branches that were overhanging, that he brushed with his head when he walked by. They were anywhere between seven and a half to eight feet off the ground. We then went slowly to the top of the hill, because we didn't want to run into it. We never saw it again."

Q: "What date did this take place?"

A: "Late October, 1974."

Q: "Can you remember the exact date of this incident?"

A: "No, I can't remember the exact date at all."

Q: "Was the sighting at night or day?"

A: "It was at 9:40 in the morning."

Q: "Was it still dark?"

A: "No, it was getting light, and it was lightly snowing at the time."

Q: "Describe the area in which this took place."

A: "The area is in Princeton, just across a wooden bridge. The town is on the south side of the bridge, and the sighting I had took place on the north side where there is a road along the river that goes east and west and you got houses spread out along the road, and it's called Allison Flats. You have a cliff here that goes, I don't know, maybe 300 feet, and there's the airport up top, just a municipal airport, eh, not much to it. It wasn't even paved at the time, I don't think."

Q: "The sighting took place on the other side of the bridge?"

A: "Yeah, just on the other side of the bridge, on the west side."

Author's note: Glen has his directions wrong here. The bridge in question lies north-south. The incident occurred on the north side of the bridge.

Q: "What was your first reaction when you saw what you saw?"

A: "Well. I was with my brother...he's really gullible so he wanted to go right after it, and I was really scared. I didn't know what to expect eh, because it was pretty big."

Q: "What was it doing?"

A: "Well, we had just gone around the airport eh, there's a circular path or road around the airport, and we always jog there in the morning. We had just come down on the east side of the road there and we were walking the last 200 feet because we were tired eh. So we are just walking and we saw something crossing the road on the west side (north side) of the bridge. So we stopped to have a look, and it only took about four steps to get across the road around the airport, and we always jog there in the morning. We had just come down on the east side of the road there and we were walking the last 200 feet because we were tired eh. So we are just walking and we saw something crossing the road on the west side (north side) of the bridge. So we stopped to have a look, and there is a little ridge there that it wanted to get up, on a dirt road that goes up and around the airport, and it only took it one step to get up there. He just lifted himself right up on that roadway and started to walk up it eh!"

Q: "Did it walk on two legs?"

A: "Two legs, yeah, even when it went up that one step, it didn't drop down on all fours like a bear, eh?"

Q: "It never dropped down on all fours at all?"

A: "No."

Q: "Was it covered with hair?"

A: "Yes it was, yes."

Q: "What color was it?"

A: "I would say brownish black."

Q: "How tall would you estimate this creature to have been?"

A: "We estimated about seven and a half to eight feet. Because when he went up the hill, later we followed, okay, when he went up the hill he was just brushing branches that were overlapping the road. Like they are just hanging off the bank side there, and they were off the road. So I couldn't touch them, eh. So I think the thing was seven and a half to eight feet tall."

Q: "What would you estimate its weight to have been?"

A: "He was heavy, for sure, and...I don't know, 400, 500 pounds. Who knows?"

Q: "Did you see any facial features?"

A: "Ahh, well I didn't have the time. Like when we stopped and watched eh, I was really terrified myself. But like I said to my brother...it had its back to us as it walked up the roadway, like on the final stretch there when you are going up the hill and...my brother yelled out at it. He said, "Hey!" And it turned around just long enough, and it was just like you were seeing, I don't know, King Kong or something. It was definitely an apelike face."

Q: "Could you describe the eyes or nose?"

A: "The nose was not protruding much and the eyes were really dark eh, but I couldn't really tell."

Q: "Ears or mouth? Did you see its teeth at all?"

A: "No. It had...I wouldn't say it had a real hairy chest, we were debating through the years whether it was male or female, because it did have all the fur around the chest area. Unless it was muscle? I don't have any idea, but it was different, like it wasn't just fur?"

Q: "Could you describe its arms?"

A: "The arms hung quite low. Past the knee area anyway, very big and covered with fur."

Q: "How long did you see this creature for?"

A: "The duration of the sighting was three minutes, if that. It wasn't worried about us at all, it didn't run, it didn't care. It didn't see us until it crossed the road and was already up on the bank, heading up for the roadway eh."

Q: "Did it ever make any noise?"

A: "No. No noise."

Q: "Did it see you?"

A: "As far as I'm concerned it didn't see us until my brother yelled at it."

Q: "What was its reaction when your brother did yell?"

A: "It might have been slightly startled. But he turned and we could see, uh, what it was like and I'm sort of glad at this time that he did because otherwise we would never have known. Cause at first, when we first saw it, we thought it was some sort of prank or whatever. Some individual with a long fur coat. But this was different, like it was clinging to him. So it wouldn't have been that, eh."

Q: "Did you smell anything during or before this sighting?"

A: "No."

Q: "After the creature moved off did you check for footprints?"

A: "We did eh. It took my brother about ten minutes to convince me to follow him, okay. He wanted to follow him right off the bat. But like I said, going up the hillside there, it was pretty brushed in there and I didn't want to get involved with it. But when he headed up around the corner, I think it was maybe ten minutes later, and we headed up that same place where it was and we had seen where he had moved the dirt and that to get up the hill, and then we got his prints, okay. My brother tried to preserve it by putting cardboard over it, there was cardboard lying around in that area, so we got some cardboard and just threw it over. Because it was lightly snow-ing, like it was just a skiff of snow. Enough for him to leave his print, and we could see that the print was bare. We knew that. My brother, like he was...I don't know how old he would have been then. He was younger than...I forget what grade he was in, seven or eight. He had some sort of measuring tape with him, a small meas-uring tape, just in his jacket pocket, that he uses on his...like they have their sort of war games, okay. I don't know if it was Dungeons and Dragons, or who knows what, but he has that for measuring stuff. We measured the print and it was about thirteen inches."

Q: "Thirteen inches long?"

A: "Yeah. Like I said, you could see the difference in like there was no shoe mark or nothing. It was definitely bare, whatever he was walking on eh."

Q: "Was the print very deep?"

A: "No. The roadway there is fairly hard, like it was getting to be frozen or whatever from upcoming winter or whatever, so it was pretty hard. But like I said, it was leaving footprints in the snow."

Q: "What is your brother's name?"

A: "Alan."

Q: "Did you reporwhat you saw to the police?"

A: "No."

Q: "Did you report it to anyone?"

A: "Just family members."

Glen Boulier, since my interview with him has moved with his family away from the house he occupied in 1986, and I have now lost contact with him. I found him to be an easygoing family man who loves his children and his wife. I don't know what else it could have been that he and his brother Alan encountered while they were out jogging in late October, 1974. But it is my belief that if the sasquatch does indeed exist, they saw one.

During the summer of 1986, I stopped by Princeton to have a look at the bridge where Glen Boulier and his brother saw a sasquatch in 1974. Of course it was twelve years later and I had no hope of seeing some leftover sign of a sasquatch walking past that day. I just wanted to take some photos of the area for the file.

Above: Glen Boulier took this photo of the bridge in Princeton two days after he and his brother encountered a sasquatch while jogging here in October, 1974.

Below: The same bridge area as it appeared in May, 1996. Progress it seems has begun to overrun this area as development crosses the Similkameen River.

What I didn't know at the time was there was much activity going on just a few miles northwest of Princeton near the small community of Coalmont, B.C. I would learn of this in 1991. It's too bad that I wasn't informed at the time I was in the neighborhood, for I would have investigated on the spot while all this was going on. Most of this activity was recorded by a man I will refer to as Barry Stanley (not his real name), who at the time lived in Princeton. Most of the sightings took place near a hunting cabin he owned just northeast of there near Coalmont. Mr. Stanley continued to correspond with sasquatch researchers until he moved away from the area to Nanaimo, B.C., in 1987. We don't hear from him anymore so I assume his interest in sasquatch research has given way to other things. Also it was my understanding that he was suffering from poor health at the time of the move, so it could be possible he may have passed away. At the time of the incidents he requested that his name not be revealed publicly, the same request was made from many of the witnesses. Since nobody has heard from Mr. Stanley since 1987, I will still honor his request in this book, as well as those of the witnesses.

In a letter dated, July 2, 1986, Mr. Stanley gave an overview of strange happenings that had occurred during the month of June, 1986, at the site of the cabins. He writes:

Location: On Granite Creek, approximately midway between Newton and Bakeburn Creeks. 5.5 miles from Coalmont, B.C. Access, good road access to within 1/8 mile of cabin.

Summary of Activity: 1st week of June. One night, the occupants of the cabin noticed that everything had become dead silent, all felt a feeling of apprehension. This was around midnight. A short time later, something jarred the cabin violently. Occupants too afraid to go outside at night to investigate. Investigation in the morning showed no evidence of tracks or anything unusual. Noises were heard in the bush awhile after the cabin was shaken that the occupants described as, "sounding like trees being broken". The exact date of this happening was not remembered.

2nd Week of June: several instances of trip wires being broken, and of noises being heard that resembled children at play, however there was nobody anywhere near the cabin or the immediate vicinity.

3rd Week of June: 2 boys saw a large black head in the bush. When the creature noticed them it fled upright down a hill. This was just around dusk.

4th Week of June: 23rd and 24th as well as the 25th were quiet.

26th of June: Two men heard strange high pitched whistles emanating from the timber. There were no campers or hikers in the area. The whistle like noise was heard five or six times.

27th of June: Tracks were found on a hillside. There were 12 tracks in line, and were measured at 17 inches long, by 6 inches wide. The tracks were impressed about 1 and 1/4" deep into dirt that the men didn't leave a track on. The stride was measured at 5.5 feet. The tracks were indistinct. However a human like shape could be seen to them. The men are experienced woodsmen and state that it was 'NOT' a bear track. The track was made by something bipedal.

Author's note: One mistake most new researchers make when dealing with footprints is confusing step with stride. Step is the distance between heel of left foot to heel of right. Stride is distance between heel of left foot to heel of left foot again, or heel of right foot to heel of right foot again. It could be that these men were measuring steps instead of strides, but of course there is no way to be sure. Also, as far as I know at this time these alleged footprints were not photographed. Mr. Stanley continues:

28th of June: At 1:30 in the morning, one of the men went outside to urinate, and while shining the lamp around, noticed eye shine of a bluish green colour on a rock bluff behind the cabin some 150 feet distant. When the light was played back on the eye-shine, a large bipedal creature was seen to run off into the dark. A short time later, some trees were shaken just below the cabin. The men state that the trees were some 6 inches in diameter. Two eyes of yellowish green shone in the light when the light was pointed over the bank to investigate.

29th of June: At about 11:30 p.m. A large rock struck the east wall of the cabin, leaving marks on the logs. The rock was found the next morning and weighed about 10 pounds.

30th of June: Trees again shaken after dark near the cabin, no wind.

1st of July: 4 a.m. A large rock was thrown at cabin again. Marks left on the logs, and rock was found, weighing about 8 pounds. 7:30 p.m. The men heard what they described as a "chattering sound", and stated that it was definitely not a squirrel. Shortly after that, they heard several, "grunts, moans, and monkey like sounds". This was about 1/8 to a mile from the cabin near a cedar swamp on Newton Creek. They observed something dark, dart behind a tree but were too afraid to venture up to see what it was. Later that evening, some indistinct noises were heard, and also sounds that sounded like, "children laughing". However there were no campers or hikers in the area.

2nd of July: 4:30 a.m. another rock was thrown at the cabin.

Author's note: Children laughing? Perhaps the men heard who was really responsible for the strange goings on. However, there were sightings of a creature as well, and more was to follow as Mr. Stanley's stepson Mike was to have a sighting himself a little after the men using the cabin left.

Mr. Stanley's stepson also requested his name not be revealed at the time. I do not know the exact date of Mr. Stanley's stepson's sighting, however it did occur between July 2 and July 8. There is another cabin, old and not used, not too far from Mr. Stanley's, where the men had all the strange happenings. Mr. Stanley on July 10 interviewed his stepson about what he had seen.

Stanley: "Mike, could you describe basically what happened for me?"

A: "Sure. I was going down to the creek. And when I passed the cabin, I saw something out of the corner of my eye, just a movement sort of."

Stanley: "The cabin? This was the old, abandoned cabin down below the weather station?"

A: "Yes. I looked up and saw this...creature hunched over or squatting on the roof of the old cabin. It was big, and it was light brown, about the color of a cinnamon bear. That's what I thought it was at first. Then it stood up and it wasn't a bear."

Stanley: "About how far away was it?"

A: "About from here to the shed across the street here" (approximately thirty to thirty-five yards).

Stanley: "Okay, can you describe it, one thing at a time? How about size?"

A: "It was hunched over, so I couldn't tell at first. When it stood up, it was...big! I couldn't tell just how big because the peak of the old roof obscured part of its legs and its feet. I would say it was maybe seven and one-half to ten feet tall, depending on just where on the roof it was standing."

Stanley: "What color was it?"

A: "It was brown, sort of a light brown."

Stanley: "Did you see the face?"

A: "Yes. The face was sort of a blackish color, and there wasn't much hair on the face. I remember it had a large mouth, but the mouth was closed. I didn't see the color of the eyes. It had no nose that I could see, looked like two hole where its nostrils should be."

Stanley: "How about the hair on the head, did you notice anything about this?"

A: "The head did have a lot of hair on it, but the hair wasn't much longer than anywhere else on its body. It was the same color as the rest."

Stanley: "What did the head look like, as to shape?"

A: "It was sort of pointed, not real pointed like a gorilla, but it had more of a point than a human. The point was sort of in the middle of the skull, not at the rear like a gorilla or ape."

Stanley: "Can you describe the body?"

A: "This thing was without a doubt female. It had two breasts, and they were long, and very droopy. The chest was massive and powerful looking."

Stanley: "You mention breasts. Did you notice if they were bare or had hair on them?"

A: "They were covered with hair about the same length as on the rest of the body."

Stanley: "Go on."

A: "The arms weren't overly long for the body. They looked pretty much like a human. But God, were they muscular! The hands were about level with the knees, not hanging below the knees like a gorilla or orang."

Stanley: "Did you see the legs?"

A: "Only partly. It was standing on the far side of the roof and the roof peak was between myself and it. The top part of the legs appeared to be extremely muscular also."

Stanley: "Did you get a good look at the hands?"

A: "No. I couldn't see them that well. Couldn't tell much about them."

Stanley: "About how long did you observe the creature?"

A: "About five to ten seconds or so."

Stanley: "What did it do?"

A: "Nothing. That's what scared me. It just stood up and looked toward me and stood there for a few seconds, then it jumped off the roof and ran into the bush."

Stanley: "Ran? How did it run?"

A: "Ran away, fast, real fast. It ran like a man, on two legs."

Stanley: "Did you notice any smell?"

A: "No, couldn't smell anything."

Stanley: "Was there any wind?"

A: "No, it was calm in the canyon."

Stanley: "What was it doing on the roof? Did you see it pick up or carry anything at all?"

A: "No, the hands were empty. It was either stooped over or it was squatting when I first saw it."

Stanley: "Did you look up on the roof later, to see if you could find hair or anything?"

A: "No. I just wanted to get out of there. I guess I really didn't believe what I had just seen at that point."

Stanley: "What did you do after you saw it?"

A: "I was a bit shook up, but I went on down to the canyon."

Stanley: "Did you find any tracks?"

A: "I looked behind the cabin where it jumped down when I was going on my way up, but the ground was real hard and I couldn't find any."

Stanley: "Can you remember anything else Mike?"

A: "No, that's pretty much it."

Stanley: "Mike could you put down a sequence of events for me on paper in your own words as to what happened? Sort of just like a statement?"

A: "Sure, no problem."

Author's note: I do not have a copy of Mike's written statement for my files, but I do have a copy of a picture he drew of the creature he saw on the cabin roof. Like I said before, at the time this was all occurring, Mike, like Mr. Stanley, did not want his identity revealed, so that is why I've only printed his first name.

There was one more reported sighting during all this, and that was by Mr. Stanley himself. He, too, reported that he saw a female sasquatch, downhill from himself behind a large rock. As did his stepson, Mr. Stanley observed that the creature had droopy hair-covered breasts. Perhaps it was the same creature? Mr. Stanley's sighting happened near Granite Creek on July 26, 1986. He, too, thought the color of the creature's hair was light brown and that it stood about seven and a half feet tall. He also drew a picture of what the thing looked like, which I now have and is included in this book. It wasn't too long after that Mr. Stanley moved away from Princeton to Nanaimo, B.C., soon after to disappear from the world of sasquatch research. Of course, with his departure we have no way of knowing if sightings continued around these cabins near Coalmont.

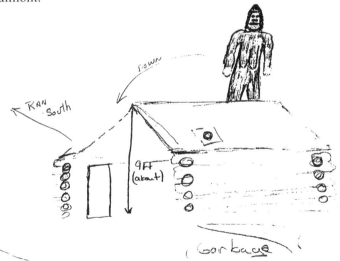

Sketch of creature seen by Barry Stanley's (not his real name) stepson, Mike. The creature was seen on the roof of an abandoned log cabin near Coalmont, B.C., in early July, 1986. The creature appeared to be a female due to the fact it had large hair-covered breasts. Sketch was drawn at Mr. Stanley's request.

97

On July 26, 1986, Barry Stanley (not his real name) saw a female sasquatch, near the cabins, and he drew this sketch. Perhaps it was the same creature his stepson Mike saw three weeks earlier. These two reports were the last of a series of events which occurred near Coalmont, B.C., during the months of June and July, 1986. I was in the area myself at the time looking into a twelve-year-old incident, and was not informed of the incidents going on at this time. I wish I had been.

There have been few incidents since the turn of the century, in which someone has reported a sasquatch being shot. One of the main points of those who doubt the creature's existence is the fact that no hunter has ever shot and killed one. At least at this point in time no hunter has come forward to tell the world about killing one. I was contacted by a man who said that as a teenager, he and two friends, while hunting small rodents with .22-caliber rifles, shot at and hit what he thought was a sasquatch, in a place called Hunter's Range, about thirty miles from Enderby, B.C., in late July, 1973.

I went to Pete Nabb's home in Calgary to interview him about what he saw that day. I found Pete to be an likeable, outdoors type who loves hunting and fishing. He told me that he has told of his encounter to others and had been ridiculed and even called a liar on one occasion. When I sat down with Pete he seemed a little nervous, and he looked down at me and said, "Look, I'm not interested in being put down, so if you're here to call me a liar, we shouldn't go any further with this." I assured him that I didn't have an opinion yet one way or the other, and suggested we get on with it. He didn't say anything for a few seconds, then he just nodded and said, "Okay."

Q: "What date was it when you saw this thing?"

A: "It was high summer, late July in 1973."

Q: "It was you and two friends who saw it?"

A: "Yeah. Me and two friends."

Q: "Where in British Columbia did this happen?"

A: "Hunter's Range."

Q: "That's about thirty miles from Okanagan Lake?"

A: "Well, it's about thirty miles from Enderby."

Q: "What time of day was this?"

A: "It was in the afternoon."

Q: "Just describe what you saw and what happened."

A: "Well, we were walking through a meadow and we got, I don't know, maybe twenty-five yards into the meadow. We were looking around, you know with a .22-cal, you hunt little things, and we saw, I'm not sure who saw it first, probably all three of us at the same time. It was standing there and it didn't move, but we saw it. At first we thought it was a bear. But it wasn't a bear, and we shot at it. We looked at it, it didn't move, it didn't run away from us or attack us or anything. We didn't walk any closer to it, we stopped where we were, and the three of us...we shot at it and we emptied most of our guns at it and nothing happened, it didn't fall, we didn't kill it or hurt it."

Q: "No reaction, it just stood there and watched you?"

A: "Yeah, it just stood there and like it didn't move; it didn't run away or nothing. And...after we shot at it and that sort of suggested that it would be a good idea to buzz off out of there in case it gets mad and comes after us or, you know, gives us some of the treatment we were giving it."

Q: "Did you three hear it first or see it first?"

A: "We saw it."

Q: "It was just standing there, watching you?"

A: "Yes, there was no noise. It didn't make any kind of noise."

Q: "You were in the meadow and it was standing in the tree line?"

A: "On the tree line, yes."

Q: "What color was it?"

A: "It was black."

Q: "Black?"

A: "Yes."

Q: "Was it covered with hair?"

A: "Yes, it was covered with hair."

Q: "Was it standing on two legs?"

A: "Yes."

Q: "You didn't notice any face features?"

A: "The face features I saw were...you know thinking back years ago, would be like, the eyes, they were deep, they weren't...they were deep eyes. And the face, it had a nose like...like a Negro's nose."

Q: "Like a flat nose?"

A: "Yes, and a like...it had a mouth, and it was big."

Q: "You didn't see its teeth?"

A: "No, I didn't see its teeth."

Q: "The arms, were they large, massive arms?"

A: "They were big arms, and they were really long."

Q: "Covered in hair?"

A: "Covered in hair, yes."

Q: "Did you notice the color of the skin in the facial area and hands?"

A: "I didn't see any skin."

Q: "No?"

A: "No."

Q: "The face was all covered with hair?"

A: "Yeah, that's all I saw was hair."

Q: "Okay, you didn't get a good look at its eyes?"

A: "No."

Q: "As soon as you three saw this thing, you started shooting at it?"

A: "Yes, we started shooting at it pretty quick. Like it was a few seconds, we had to decide what it was so...we thought it was bear. It wasn't a bear, because bears don't lean against trees on its hind legs, and this thing was leaning on some trees, standing up, and it was bigger than a bear."

Q: "When you shot at it, you didn't hang around to check for footprints?"

A: "No we didn't. We were just kids then, and you know, it scared us because you know when you have a gun...you're told with a gun, you shoot something, it's supposed to die. And this thing didn't die, and after we shot at it, we turned around. Like, I suggested that we leave. And we turned around and we started heading back, and we were cautious on...we didn't want to get hurt, and we all turned around and we looked, and we saw it turn. It quickly turned, for something that big it...like it turned...fast, and it walked off into the bush. It wasn't running, but it moved pretty fast, and it didn't back or come after us or nothing."

Q: "It walked on two legs at all times? Never went down on all fours?"

A: "Never."

Q: "How tall would you say this creature was?"

A: "I would say that it was eight and a half to nine and a half feet tall."

Q: "Nine and a half feet tall?"

A: "Yes."

Q: "You did not go back to check for footprints?"

A: "No we didn't. It was pretty scary. When it walked away, like when it turned and it went like a real heavy human, like a real heavy man."

Q: "Did you smell any odor?"

A: "No, we were too far away to smell it."

Q: "It never made any noise?"

A: "It never made any noise at all."

Q: "There was no violent actions by this creature?"

A: "No, not like we were giving it."

Q: "You shot at it and the bullets had no effect?"

A: "The bullets had no effect, and I know we hit it! The three of us were not bad shots, and I know that some of those bullets hit it, and it didn't, you know. If I had been older, I would have gone over and checked things out, you know, more closely. Maybe even followed it."

Q: "How old were you at the time this happened?"

A: "Well I'm twenty-seven now, so I would be thirteen or fourteen."

Author's note: I interviewed Pete Nabb in 1987. He would be thirty-seven at the time of writing this book.

Q: "It just moved off when you three boys ran away?"

A: "Yeah, when we turned and started to leave, it did the same thing. It's strange, I never saw anything like it in my life."

Q: "You haven't seen anything like it since?"

A: "Nope."

I know a .22-caliber rifle would not be powerful enough to bring down a large animal like a sasquatch; however, I do have to believe they would still have to hurt pretty bad. So I can't imagine such a creature just standing there while bullets were striking it. I suppose in their fear they may have been missing the target. I talked with two of Peter's friends from work who showed up at his house as I was leaving his apartment. They told me that Pete is no-nonsense kind of guy who likes camping and hunting, as well as fishing in his spare time.

Pete did make a mistake during the interview when he said bears don't lean upright against trees. He is wrong here. Both black and grizzly bears have been known to do this. However, bears do not walk for long distances on their hind legs. The creature these boys saw apparently could and did. So what was it these boys shot at in 1973—imagination, a bear or a sasquatch?

Reports from this region of B.C. (near Okanagan Lake) have been few and far between; I do not know why. Yet, if one goes just an hour or so west, near Princeton, reports climb considerably. The same holds true a few hours to the east, in the Kootenay region. Why this dry spot in between, I have no idea. It could be, like Vancouver Island, that I have just not heard about them.

I did receive a phone call from a Mr. Robert Ebb of Vancouver, on September 24, 1997, who wanted to tell about strange footprints he found while hiking in Okanagan Mountain Park on the southeast shore of Okanagan Lake, during the fall of 1995. He told me that he had been walking along an abandoned rail bed, when he came across two footprints, imbedded three inches into the hard dirt, where he himself left no impression at all. There were two prints, one left foot and one right. He guesses that they were at least sixteen inches long, though he was not sure. He did not say that they were bigger than his own size-twelve foot.

He did not photograph the prints at the time since he was not carrying a camera. He did tell me that he would make sketches of how they appeared and send them to me in the mail. At the time of writing, they still have not arrived. Then he told me something that pretty well had me thinking that somebody was playing games with him. The prints had six toes rather than five. The sasquatch, if it does indeed exist, has five toes on each foot, not three, not six, but five. I have heard of such strange tracks being found in the eastern U.S. but not in British Columbia and Alberta. When I hear of footprints with anything other than five toes, I dismiss it out of hand as either an attempted hoax, or a case of mistaken identity. I don't think Robert Ebb was trying to attempt a hoax, for he was asking me about the toes and if I had in my possession any other reports from my files which contained six-toed prints. I told him no, at least not any I would consider to be authentic.

Since I began this chapter talking of happenings around Harrison Hot Springs, B.C., I think it would be appropriate to end this chapter with another report that occurred near this community. Sasquatch Provincial Park lies just northeast of Harrison Hot Springs. And as you can guess by the park's name, the fact the sasquatch has been reported to wander the mountains around Harrison Lake helped the B.C. government choose a name. There had been reports from the area before it became a provincial park. As well, there have been some since. The park consists of three main campgrounds: Hicks Lake campground, Deer Lake Bench campground as well as the Deer Lake lakeside campground. All these campgrounds provide outhouses, picnic tables and fire pits for visitors. Fire wood is stacked in bins and campers may take all they can burn. The area is beautiful and I myself have enjoyed staying there many times since the early 1980s. However, being beautiful and popular has its drawbacks. The campgrounds on all long weekends are almost always full. So go early or forget it!

In August of 1994, it seems that the park's namesake had decided to pay a visit and a couple of sightings were reported, as were strange screaming noises coming from the hills at night. It was reported that several camping families packed up and left, though I have no direct confirmation of this. I learned of the happenings in Sasquatch Provincial Park when I received phone calls from two different people on the night of August 17, 1994. Neither one of the

103

Above: Highway #7, which runs between Vancouver and Hope, B.C. is officially named Lougheed Highway. Unofficially it has been called Sasquatch Drive. Many of these signs were seen along its length. In the last five years though many of them have begun to disappear. Removed for road work, and not put back after the work has been completed. I photographed this sign in 1993. It has since been removed.

Below: The Sasquatch Inn, which is located along Highway #7 just west of Harrison Mills, B.C. An old landmark which will hopefully remain for some time to come.

callers knew that someone else had called to report the same incident. The first call was from a Mr. M. Jasmin of New Westminster, B.C. Mr. Jasmin was camping in site twelve at the Deer Lake lakeside campground, with his family. In the back of site twelve, there is a path that Mr. Jasmin thought might lead to the lake, so he decided to follow it. About twenty-five feet down the path he found what he thought to be four footprints. He told me he would send a written report on what he found, and true to his word he did.

The second caller was a Ms. Sylvia Pool, who works as a fee collector in the campgrounds and also saw the tracks. She was doing her job with her twenty-one-year-old daughter Veronica keeping her company, when Mr. Jasmin waved her down to show her the prints. She did not remember Mr. Jasmin's name, but referred to him as "the nice man in site #12." I did not tell her that Mr. Jasmin had already called to tell me about his find, about two hours before she did.

They both told me that one of the prints was clear and distinct, showing all five toes clearly. Mr. Jasmin informed me that he staked one of the tracks off with small sticks. Ms. Pool's daughter, just before she hung up the phone, told me in fact only the big toe was clear and also the heel. Mr. Jasmin maintained that he could make out five toes. Also, Veronica informed me that her mother's boss had a look at the prints and found them interesting, and that he was going to inform his superiors about what was going on.

I called John Green in Harrison Hot Springs that same evening to see if he could go out there right away and have a look. He and the late Robert Titmus did just that. Later that night John called me back to tell me of his findings. The Jasmins had left the site and it was now unoccupied. This I already knew as Jasmin had phoned me from his home in New Westminster. What John and Robert found though was not encouraging. Essentially, the alleged footprints had been tramped upon by people and were now nothing more than shapeless depressions in the ground, including the one which appeared to have been staked off. Neither Mr. Jasmin nor the women had photographed the prints, so we have no idea if the detail was any better in the first place. However, when Mr. Jasmin's written report arrived about a week later, I was left with the impression that indeed they were. He writes:

Tom:

You may be interested to read this report which, incidently coincides with a "Sighting" of a Sasquatch near Harrison Hot Springs. The sighting apparently took place at Bench Provincial Park last week.

Author's note: He means Deer Lake Bench campground. Some people refer to the campgrounds as parks themselves. He continues.

The incident was accompanied by the detection of a strong odour, according to Park Rangers. I can only relate this second hand because I was not present when it happened. Also, according to park authorities, the news of a sighting caused a state of panic in some campers who quickly folded up and left the campsite. My experience is quite different but just as exciting. I was camping with my family at Deer Lake Provincial Park, (about 3km from Bench Park),

Author's note: Again he means Deer Lake lakeside campground, as well as Deer Lake Bench campground. He continues.

When I noticed a small trail leading from our campsite, (no.#12), toward the mountain. I thought it may lead to the lake and proceeded to follow the trail. About 25 feet from the entrance, I noticed what I thought was a deer track.

Being a hunter, I found it strange that there would be only one such print. I looked at the trail closely and noticed that the ground had been disturbed and seemed to have slid from the light grade of the trail. That's when I noticed a clear, well imprinted track, (footprint), in all fashion indentical to those shown on T.V. programs about Sasquatches. I stopped and looked and began to see other footprints leading to that one. In all, four prints. One I unfortunately damaged by walking on it. I immediately advised the park authorities who dispatched the supervisor.

Author's note: I assume, when he says he advised park authorities, that's when he waved down Sylvia and Veronica as they were doing the rounds collecting fees. Remember, he did not know that

106

they too had phoned me the same evening to report the same thing. Also they did not know that he had called. He continues.

I marked the print with sticks by forming a square around each of them. The park supervisor contacted the B.C. Parks Department who dispatched someone due to arrive today, August 17th.

Author's note: I wonder if instead of calling the parks department they ended up phoning me? I was not informed of this going any higher or anyone from the Parks Department coming out to investigate. But perhaps they were referring to the woman's boss who came to have a look. As far as I know, John Green and the late Robert Titmus were the only other people to go to the site at my request on August 17. John did tell me that the alleged prints had been badly stamped upon when they saw them. Maybe park authorities did go to the site and did a little more than just have a look? He continues.

Since we had to get back to Vancouver, I spoke to park authorities and told them to guide the man to the tracks. I don't know what has happened since then but you may be able to get some information by contacting B.C. Parks and finding out who was on duty during the day of August 16th and 17th. Before I left, I made sure to get as much information as possible about the footprints. I hope this may be of interest to you.

Data on Left Foot.

A very clear, well defined print showing the heel, all the toes and the arch of the foot. (Note: all dimensions taken as outside dimensions).

- Dimension from big toe to heel: 15-1/2"
- Dimension from small toe to heel: 14-1/2"
- Overall width at toes: 9-1/2"
- Width of the heel: 5"
- Depth of the heel: apx. 1-1/2"
- Depth of the small toe: apx. 1/2"
- Depth of big toe: apx. 1"
- Stride: apx. 3-1/2 feet.
- Feet apart, (Step), apx. 2 feet.
- Soil quality on August 16th: Top soil, dark, very fertile, no rain for 42 days.
- Estimated weight of animal between 400 and 600 lbs.
- Tracks found on a slight up grade.

- No accurate data on other tracks because of foliage in the way, however, clear enough to be recognized as belonging to the same animal.

Good luck.

Peter M. Jasmin

Clearly Mr. Jasmin paid close attention to the prints he found as well as noting details carefully. That is why I'm confused as to why John Green and the late Robert Titmus reported only shapeless depressions in the ground when they went there later on the night of August 17. I don't want to make any accusations here, but I have to wonder what happened to them between the time Mr. Jasmin left site #12 and John and Robert arrived

John Green, seen here high above Harrison Lake. John has been investigating reports of sasquatch since the mid-1950s, and has written extensively about them. He and the late Robert Titmus went out to investigate the tracks reported by Peter M. Jasmin, at my request. He reported nothing but shapeless depressions in the ground, quite different from the detailed footprints Mr. Jasmin reported. Again I have to wonder what happened between the time Mr. Jasmin left site #12 and when John and Robert arrived.

Photo: T. Steenburg, 1997.

Above: Site #12 at the Deer Lake Bench campground, Sasquatch Provincial Park. Occupied by Peter M. Jasmin, August 16, 1994, who phoned me to report he had found footprints along a small path behind the campsite (right). Two park employees also reported the tracks. I contacted John Green who, with the late Robert Titmus, went to investigate and found that the alleged tracks may have been obliterated by persons unknown before they could be studied.

Photos: T. Steenburg, 1995.

British Columbia
Statistics

Since researchers started hunting the sasquatch in the mid-1950s, many have tried to find patterns in the creature's behavior in order to predict its movements. Over the last forty years though, it has become obvious that we really can't predict where the sasquatch might turn up at any time of the year. We don't know if sasquatch migrate, we don't know if they have a mating season, we don't know if they follow any particular growing season. The more we try to find out, the more we realize that we don't know anything about their habits, feeding patterns or reproductive patterns.

The purpose of this chapter is simply to point out some characteristics of the animal, based on witness observations. These statistics are based solely on the information in my files from the witnesses I've interviewed and they only refer to sightings in B.C. They do not necessarily reflect the sasquatch's true character. One thing some will notice, is that my statistics contradict some accepted behavior traits. It is commonly believed that the sasquatch is mainly a nocturnal creature. Many well-written books on this subject back this statement up, and they may be right. All I'm saying is that the statistics from my files contradict this. However, it is indeed possible that in future interviews these statistics may change.

Many other books on this subject have suggested that the most likely time to observe a sasquatch is during the fall. According to my files, summer reports outnumber fall reports by a fair number, at least in British Columbia, which is likely due to the fact that people camp and spend more time outside in the summer. I have twenty-seven reports from the summer months or 32 percent of the total so far collected. Nineteen reports (21 percent) occurred during the fall. Spring follows with thirteen reports (14 percent). Winter, as expected, is last, though not by much with twelve reports (13 percent). Much of the seasonal stats are open to change. And they are based on the correct dates of season change. Many people are surprised when they realize that most of December is officially considered fall. Winter actually begins on December 21. Also, I have sixteen reports in which the particular season was not given (18 percent of the total). If the actual dates were known, I'm sure many of these statistics would change.

As I said before, my statistics do not support the claim that the sasquatch is a nocturnal creature. According to my files, most encounters have occurred during daylight hours. Sixty-four reports happened during the day, or before 6 P.M. (77 percent of the total). Nineteen occurred at night, or after 6 P.M. (23 percent). The sun stays up well past 10 P.M. during the midsummer months in most of Canada. In midwinter, it is dark just after 4 P.M. This difference of course is due to our northern location.

The gender of the creature seen is most often not noticed by the person who sees it. This is understandable since most encounters are very brief, lasting a few moments at most. Most witnesses who claim the creature was female have done so due to the fact the creature they saw had mammary glands. Those who say it was male base this on the fact it didn't have mammary glands and was flat chested. Very few witnesses have reported seeing male genitalia. Out of the total number of B.C. reports in my files, seven reported that the creature was male (7 percent). Six reported that the creature was female (6 percent). The vast majority had no idea as to the sex of the creature they were looking at in seventy-three reports (87 percent).

Hair color is almost always brown or black. I have twenty-one reports where the hair color was reported to be brown (24 percent). The second most common hair color is black according to fifteen reports (18 percent). A good many witnesses really didn't know the hair color, so they just described it as dark in twelve reports (14 percent). I have seven reports stating the color was reddish brown (7 percent). Three reports talk of a gray-colored creature (4 percent) and another three reports have a creature with black hair with gray tips (4 percent). Another two reports have the witnesses saying the creature was blackish brown or brownish black (3 percent). I have only one report where the creature was described as light colored (1 percent). Also, I have twenty-two reports in which the color was not given (25 percent).

If anything has been established, it is the fact this creature is physically huge and manlike in its locomotion. There the similarities end. Such a creature would have to have developed a means for dealing with cold weather. Humans deal with it by using our brains. We build houses. We have invented warm clothing. A bipedal creature without such intelligence would have only one other option. It would have to develop a body coating of hair and become very large. Most anthropologists agree that if such a creature exists, it is no surprise that it is physically huge in stature. If people were reporting something the size of a spider monkey, walking about in the forests of western Canada on two legs, there would be something wrong in an evolutionary sense. So it is really no surprise that the other biped, other than humans, on the North American continent, is apparently very large. According to my files, the most common height of a sasquatch is seven feet. I have seventeen reports of a seven-foot creature (20 percent). The second most common height reported is eight feet, of which I have thirteen reports (14 percent). Nine reports mention six-foot creatures (9 percent). Five reports mention five-foot creatures (5 percent). I have three reports of creatures standing between seven and a half and eight feet (3 percent). One report has the creature between eight and a half and nine feet tall (1 percent). Three reports mention nine-foot creatures (3 percent). One report has a ten-foot creature (1 percent) and one report has a twelve-foot creature (1 percent). I have fifteen reports in which the witness didn't know the height, but described it as very large (18 percent). In one report, the witness just said the creature was tall (1 percent). Another twenty reports had no indication of the creature's height (24 percent).

When more than one witness sees a crime being committed, in legal terms this is known as corroboration. Our legal system puts great value on this. However, the scientific community puts no value on it at all. I've heard it said that if the sasquatch story had been a murder case, we would have had a verdict and a new official species on the books by now. Most British Columbia reports in my files were single-witness sightings. I have forty-two reports of one person claiming to have seen a sasquatch (50 percent of the total). Twenty-six reports state two witnesses were present (32 percent).

Six reports had three people see the creature (6 percent). There are five reports with four witnesses (5 percent) and one report with five witnesses (1 percent). Five reports had more than one witness, though the exact number was not given (5 percent). One report had seven people all seeing the same thing (1 percent).

It would be very difficult for a man on trial to get away with it if seven witnesses were testifying they saw him do it, or five or four for that matter. It seems at times sasquatch must have the best lawyer in the continent working for them. These are just some of the statistics I've chosen from interviews I've had with people who claim to have seen sasquatch in British Columbia. We could go on forever with graphs and charts comparing everything from sounds the creature makes, to food it likes to eat. However, I am sure many of these statistics will change over the coming years as more information becomes available. None of this information is intended to give the reader a description of the sasquatch's habits or characteristics. The material simply show some details of what people have been reporting to me since 1979.

Hoaxes and the
Lunatic Fringe

If there is one thing I've learned since I started looking into this sasquatch business in 1979, it's the fact some people love to make up stories about monsters. I would estimate, since I've never attempted to keep track, for every phone call I get from somebody who really believes they may have seen something interesting, I'll get three attempted hoaxes. There have been enough books and television programs and other media outlets that anybody who does his homework could make up a story and sound completely convincing. Also individuals with nothing better to do with their time have phoned me and have been really nasty. There was a newspaper article about the forum held during Sasquatch Daze 1996, in which the reporter quoted me saying a dead sasquatch would have to be brought in before the scientific community would accept the creature's existence. (I still think this is true by the way.) This woman in Victoria took offence to my statements and phoned to chew me out for a while.

My companion Barbara answered the phone and took the full force of this strange woman's wrath. She told Barbara exactly what she thought of me and my research and what she thought should be done with me. Then she made the statement, "The sasquatch doesn't exist anyway!" Barbara, stunned by what she had just heard, replied, "Well if it doesn't exist, what are you so worried about?" The woman just rambled some more and then hung up on her.

This kind of thing comes with the territory I'm afraid, and it's something I feel I should warn anybody reading this book who thinks they might be interested in starting up their own research into the sasquatch question. You will get a lot of strange people phoning you. There's no getting around it. Another gent phoned me once and said, "I saw a sasquatch!"

"When and where did you see this creature," I asked.

"Well right now. He's sitting here having a beer with me, want to talk to him?"

No doubt they were drinking Kokanee beer at the time and a little too much at that!

Not all crank calls are people trying to be nasty. Some people have a great sense of humor and phone me as a joke. In 1991, a radio station from Bismarck, North Dakota, kept leaving strange messages on my answering machine, trying to get an interview. They always seemed to phone when I was out. The two DJs would put on strange voices, no doubt on the air, until I answered the phone. At this time George Bush was still in the White House, and when I played my answering machine I heard one of the DJs give a pretty good impression of the president's voice.

"Hello Tom, George Bush here, got a problem here with this bigfoot thing. Lay off! We know about it. We don't want you uncovering things. This is the government. We know what's happening, lay off! Word to wise, don't be messing with it. I know you're in Canada, but we can get ya!"

"Actually Tom, this is PMS again, in Bismarck, North Dakota, Cape Fire Radio. When are you ever home?" asked the first DJ.

"Yah, we keep trying, we want to talk to you about bigfoot, and we just can't seem to get a hold of you. We're going to call tomorrow at this time, tomorrow morning.

"Yah, we're your worst nightmare, we're going to call until we get a hold of you, at the same time, every single day for the rest of your life!"

"We're worried that maybe bigfoot might have gotten you."

"Yah and we may send up some mounties after you, if you don't start answering your phone."

"Right, or even some guys on horses."

"Yah. So, Tom, have a nice day, and hi to bigfoot for us. Bye."

As you can see, you must have a sense of humor in this field. If you don't, people having a little bit of fun are going to drive you crazy. The station did call again and left an even funnier message the next day, so I decided to give them the interview they were after.

Another type of caller you expect is the person who really believes in the nonsense he's telling you. These people have been politely referred to by researchers as the "lunatic fringe." This term applies to witnesses who report a sasquatch, but connect the creature to UFOs, other dimensions, the supernatural and all other kinds of weird and wonderful things. The same term is used for researchers who believe the sasquatch is part of the supernatural world. Of course, I realize that people who are skeptical about the creature's existence put all of us into this category. But in the sasquatch field, researchers tend to fall into one of two camps. One group believes the creature is just another undiscovered animal and should be looked at from a zoological perspective (I fall into this category). And then there are researchers who tend to lean toward some unearthly explanation—the lunatic fringe or the "tabloid researcher" (another term I've used over the years). This type of researcher tends to get a lot of attention from the popular press because after all, selling papers is big business and the wilder the story the better. One of the negative side effects of the lunatic fringe is that they have trivialized the study of sasquatch and thereby limited the participation of many scientific organizations and individuals because these people have reputations to protect. We have to put up with the fringe because I don't think they will go away.

Witnesses who claim lunatic fringe encounters call me all the time. During the mid-1980s a fellow, whose name I will not mention because I know he would want me to, would phone a couple times a week claiming to communicate through ESP with a family of sasquatch in the mountains around Banff, Alberta. He also claimed to have spent many days living with the creatures. When I asked for the photos he must have taken of the creatures, he replied the sasquatch had mind powers that would destroy the film in his camera, so photographing them was impossible. When I suggested he take me to see for myself, he told me that they had the ability to make themselves invisible and they only trusted him, due to his peaceful nature. It has been several years now since this fellow last called me and I hope he finally lost interest in wasting my time. Unfortunately, he is not the only one. I've received calls from—people who make this last fellow seem reasonable.

Another thing "straight researchers," for lack of a better word, have to deal with is the hoaxer. These are people who enjoy making false reports to see if they can fool you. Some have been successful over the years. There is one case of hoaxing that received nationwide publicity. It happened near Mission, B.C., during the spring of 1977. This was just before I started research myself, so I did not look into the case; however, I was impressed with the planning and its execution.

The prank involved four high-spirited young men with a sense of humor, one gorilla costume, two walkie-talkies, a fake car breakdown, as well as a half dozen or so victims riding a bus from Harrison Hot Springs to Vancouver. The bus was driven by a man named Pat Lindquist. He had driven the route many times before. The hoaxers set up their prank along Highway 7 at a point where the highway takes a long bend by a riverbank. The hoaxers even produced several fake footprints along the riverbank near where the prank was to be carried out. One of the hoaxers had his car parked along the side of the road with its hood up, thus making it look like he had a breakdown. He also had one of the walkie-talkies to tell his friends the bus was coming. Another two were in position up ahead on the highway with the other walkie-talkie, one dressed in a gorilla costume. His job was to run across the highway in front of the oncoming bus. The fourth hoaxer was on the bus in a front seat; his job was to raise hell when he saw his friend in the costume run across the road. At first everything went better than they planned. The bus went by the car, the costumed hoaxer ran across the road. However, they didn't plan on the driver stopping the bus to give chase on foot. The hoaxer on the bus didn't have to raise hell—the driver did it for him, stopping the bus and running after the costumed hoaxer. Later as the media circus started up about this sighting of the decade, Pat Lindquist was quoted describing the creature as standing seven feet tall with flaring nostrils, and it smelled terrible, giving off an odor like rotten eggs. This amused the hoaxers to no end as they had made no attempt to fake an odor at all.

Not all of the passengers on the bus were fooled though. Two Calgarians who were on the bus, Mike and Cathy Byrnes told the *Calgary Herald* newspaper the creature they saw was an obvious hoax and they got off the bus at the time and yelled to the hoaxer in the costume, "You're not fooling us!" Mike later told the press, "It just looked like a man dressed in a black monkey suit or something like that." Later when the Byrnes returned home and saw the story on the television, as well as the newspapers describing how everybody on the bus saw the same thing, they came forward to set the record straight. However, at the time nobody seemed interested in what they had to say.

The RCMP had staked off the area where the tracks were laid, to keep people away until they could be looked at. Three veteran researchers, Rene Dahinden, John Green and Dennis Gates arrived to check out the prints. At first they thought they were on to something, then after closer examination they realized the prints were the result of vertical pressure instead of a flexible heel to toe movement. Later when John was on a radio talk show discussing the case he made their suspicions about hoaxing known. During the same show one of the hoaxers called in and confessed. So the sighting of the decade became the hoax of the decade; the police stated no charges would be laid.

As I said, this was two years before I started looking into reports myself so I did not investigate. However, I do remember seeing this story on the CBC news at the time and being very excited about it. I also remember how disappointed I was later when it was revealed the whole thing was a well-setup prank. I have studied photos of the costume that was used by these four pranksters, and I personally cannot see how anybody could have been fooled by it. However we have to face the fact, sometimes people will see what they want to see. I did get the impression the bus driver had an interest in sasquatch before this happened to him. It could very well be that the excitement of seeing something he never thought he would, played havoc with his reasoning at the time. In his mind he really believed he was chasing a seven-foot, dark, hair-covered creature that gave off a pungent odor and had flaring nostrils, rather than a sweating five-foot-seven man wearing a hot, bad-fitting gorilla costume, who panicked when he looked back to see this man jump off the bus and chase him.

I received another call from a woman a few years ago who, for whatever reason, decided to try to convince me a sasquatch had walked in her backyard near Hope, B.C. It was apparent to me that the woman on the other end of the line was either drunk or high on something. She proceeded to tell me that the creature had glowing red eyes and gave off a mist from its body as it moved. She also told me that the creature was using ESP to communicate with her and it wanted to mate with her. I didn't ask if she accepted the invitation, but I did do something that I normally wouldn't do. Usually when it's obvious the person is reporting something untrue, I will politely listen to the story until they finish, say goodbye and have a laugh about it after I've hung up the phone. This time I told the woman that she was lying to me, and I didn't believe a word she said. She became hostile at this point and said, "Well to hell with you and your research!" She said a few other things which I won't repeat here, and then she hung up. I think in future I will continue to be polite until the person hangs up.

I received another attempted hoax from a man who claimed to be a game warden with Alberta Fish and Wildlife. Mr. Darren Bureshailo, of Edmonton told me that a couple from Rocky Mountain House, Alberta, named Johnston was snowshoeing near Mount Robson, when they saw a large, black hair-covered creature walk out of the tree cover, look at them for a moment, then turn around and disappear back into the trees. The Johnstons then returned to their parked car and called the RCMP. Two Mounties (identity unknown) then went to the spot with the Johnstons and footprints were found. The Mounties then told the Johnstons that this was not really a police matter, but they would call Alberta Fish and Wildlife to report this, and thus, Game Warden Darren Bureshailo went to investigate. He told me that he interviewed the Johnstons and covered the tracks with plastic, and asked that I get out there right away and have a look myself.

Well, I thought it was strange Alberta Fish and Wildlife would be called in the first place due to the fact Mount Robson is located just inside British Columbia, not Alberta. When I asked him about this, he seemed to be scrambling for an answer. Finally he told me they were closer, that's why they were called. He then again told me to go out there right away and have a look for myself. If I had received this report back in the late 1970s or early 1980s, I probably would have jumped in my Land Rover and headed out with no hesitation. When somebody yelled sasquatch I was gone. Not anymore!

What I did do was contact Alberta Fish and Wildlife to find out if Darren Bureshailo was indeed a warden in their employ. They never heard of the man. Also the RCMP did not receive any report of a sasquatch, January 24, 1995, near Mount Robson. Mr. Bureshailo does live in Edmonton and he gave me his real phone number. Not a very smart hoaxer I guess. I called a few days later and confronted him with my findings. I can only guess he thought I wouldn't bother to check him out. He just babbled something about seeing a sasquatch himself near Cold Lake, Alberta, back in 1984. Later I learned that I was not the only researcher he called. Ray Crowe of the Western Bigfoot Society had also heard from him. Ray, like myself, also contacted Alberta Fish and Wildlife and was told this man was not in their employ. It was Mr. Bureshailo's claim to be a warden that gave his attempted hoax away quickly. It did seem he did some homework on reported sasquatch sightings, since the story on its own sounded pretty good. Why he thought it would be funny to have me run out to Mount Robson on a wild goose chase, I will never understand.

As I said before, hoaxers and the lunatic fringe are part of the game, and come with the territory. They have to be tolerated. However, they don't have to be encouraged. And when I hear of someone making wild claims on some radio or television show, about how he knows the sasquatch are extraterrestrials and the rest of us are all fools for not seeing this, I will at least write a letter to the program involved informing them the difference between lunatic fringe research and straight research. I leave it up to them as to whether or not they wish to pursue the matter further. Again, I know skeptics of the creature's existence would consider straight researchers no better than the lunatic fringe. After all we are devoting a huge part of our lives to chasing down a creature that, as far as they are concerned, does not exist. If in my lifetime the impossible happens and it turns out the lunatic fringe is right, and it turns out the sasquatch is indeed an extraterrestrial or some supernatural being I will be the first to apologize. Until then, I wish they would all go away.

Mistaken Identity
and Other Errors

Unlike the hoax and lunatic fringe reports in the last chapter, there are other false reports that don't bother me at all. These are cases of mistaken identity involving people who genuinely believe they have seen or found something of interest. But after investigation, the common object or animal they mistook for a sasquatch is revealed. I always suspect this in cases where a witness didn't see the creature for a long enough period of time to actually observe it doing anything. Many of the cases you have read about in this book so far, with the exception of chapter seven, do not fall into this category. After all, if a witness watches a creature walk for any great distance on two legs, this eliminates, in my opinion, any chance of some stationary object having been mistakenly identified as a sasquatch. It also eliminates any four-legged creature being mistaken for a sasquatch. The only other possibility is the witness may have mistaken a person for a sasquatch or else the sasquatch does not exist—in which case everybody in this book is lying and this whole book should be under the heading of chapter seven. I personally don't believe this thought or I wouldn't be researching reports.

The same holds true for footprint findings. It is amazing how many people have found shapeless depressions in the ground but think they have really found sasquatch tracks. I've already written about one of these cases in chapter five, the Christopher Crocker print found near Rineheart Lake, Vancouver Island. Another case of mistaken identity as far as footprints are concerned was sent to me by a man from Rossland, B.C. He does not wish for his name to be revealed, so I will call him Elmore Butte. Mr. Butte truly believed he found something of interest, and judging from the photos he sent, I can see why. While I appreciate the photos and the report he sent, as far as I am concerned it wasn't a sasquatch track. Mr. Butte has enjoyed wildlife photography as a hobby for many years, and while he was trying to photograph bighorn sheep in Banff National Park he came across this impression in a patch of unmelted snow. His written report he sent me is as follows.

Information About Track in Snow
1. Date: March 20, 1987.
2. Time: 11 a.m. to 1 p.m.
3. Weather: clear.
4. Temperature: night -10º to -12º, day 1º to 4º.
5. Place: Banff Park, west of Banff on 1A road along Bow River.
6. Closest land mark: Sawback Creek.
7. Track found: About 1 km up Sawback Creek from 1A road, and km to km east of creek.
8. I think the track is 3 days to 10 days old.
9. The knife by the track is 3 3/8" long.
10. I calculate the track to be about 13" by 5".
11. I believe the track is Bigfoot, until someone suggests a better idea.
12. I only found this one track on the edge of a patch of unmelted snow that was about 6' to 8' in diameter.
13. I didn't follow in the direction of the track. I regret it now.
14. You have permission to use this photograph and material if you want.
15. You may not use my name in any magazine or article.
Author's note: I suspect he means any book as well, so that's why I've given him an assumed name.
16. I was alone, and my reason for being there was photographing Big Horn Sheep.

This concluded his report. When I examined the photos, I concluded that although the depression did seem to have the general shape of a sasquatch footprint, as far as I'm concerned what he found was a spot where one of the sheep he was tracking had pawed through the snow in order to eat the grass underneath, either that or an elk or deer pawed at the snow to eat. It was just a coincidence that the mark left behind resembled a footprint. I phoned Elmore Butte to tell him of my conclusions. He agreed that this could be the case. However, I was grateful to him for sending me his find in order to further investigate.

Another report of footprints came to my attention when I received a phone call from a charming couple in 1988. I have writ-

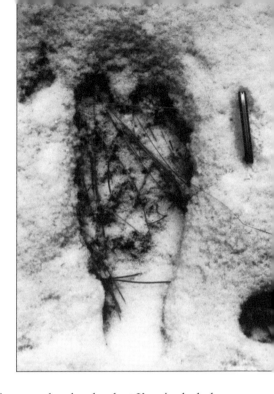

Elmor Butte (not his real name) sent this photo of what he thought might be a Sasquatch footprint, near Sawback Creek, in Banff National Park, March 20, 1987. On examination, I've concluded that this impression was the result of a bighorn sheep, elk or deer, pawing at the snow in order to eat the grass underneath. It is a good example, how people can be fooled in the wild.

Photo: Elmer Butte, 1987.

ten about these tracks before in my other books, but I've included them here because I think it's a good example of how honest people who think they have found sasquatch tracks really found something else. On the night of January 25, I received a phone call from a Mrs. Sharon Smith (not her real name) who, with her husband, found tracks that they believed were made by a sasquatch, near Abraham Lake, Alberta, two years earlier. The Smiths were hiking in early February, 1986, when they came upon a line of tracks on a frozen pond. The Smiths took six photos in all, some comparing Sharon's own size-seven shoe to the prints on the ice. There has been a long history of reports around this area through the years, and this is another reason the Smiths thought they may have found something important.

I went to their home to study the photos, which they gave me for my files. I concluded they had photographed human boot prints that were a few days old and had melted out due to warm weather. If you look closely, you can make out what appear to be normal-size boot prints inside the larger impressions. This, in my opinion, was the original size of the tracks and warm weather over a couple of days melted them so they appear to be abnormally large prints. The Smiths were a little embarrassed about having me come over for nothing. I assured them there was no reason to feel this way and they should congratulate themselves for coming forward in order to find out. I wish more people would come forward as they had.

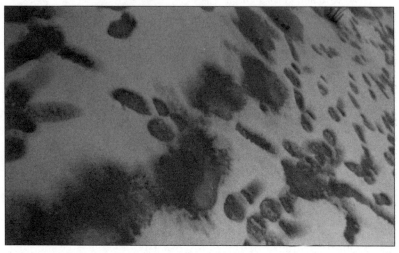

Above and below: Two of a series of six photos taken by Mr. and Mrs. Smith (not their real names) near Abraham Lake, Alberta, February, 1986. The normal boot prints you see all around the tracks are the Smiths. I've concluded that these prints are human boot prints which have been melted out in mild weather. You can see what I believe to be the original boot prints inside the larger impressions.

Photos: Sharon Smith, 1986.

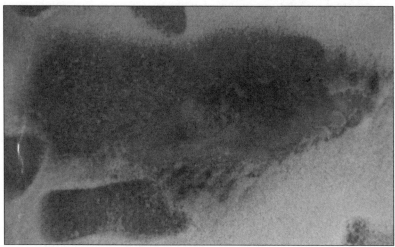

124

A researcher from Manitoba, Guy Phillips, who sent me the information concerning the sighting at Scotts Falls that I covered in chapter five, sent me two photographs he thought might be a lone sasquatch footprint. He and a friend named Bill had found it in a wooded area northeast of Winnipeg on June 18, 1988. There have been reports from this area and Guy had been investigating them for sometime. He sent me the photos to get my opinion on them. Even though the general shape was somewhat close to what a sasquatch print is reported to look like, it more resembled a swim fin people wear when they are scuba diving. I'm not saying that someone was walking around the woods of Manitoba in swim fins. But I didn't think it was a sasquatch footprint either.

What I believe happened is Guy found a place where a deer had been and the track was the result of several deer prints all mashed together. Guy told me that indeed there were deer tracks all around this area. You can see at the top of the footprint what is indeed a deer

Guy Phillips, a researcher from Manitoba, sent me these two photos of what he thought might be a lone sasquatch footprint, he and a friend named Bill found north east of Winnipeg, on June 18, 1988, in a wooded area where a number of sightings had been reported. As far as I'm concerned what we have here are a number of deer tracks all massed together. You can see what is a clear deer track at the top of the alleged Sasquatch print. Also another deer print at the top of the photo.

Photos: Guy Phillips, 1988.

print. Also, at the top of the picture slightly upper left of the track, there is another apparent deer print. I suppose it is possible that deer had walked into the track destroying some of the detail later on— but I don't think so. I also showed the Phillips photos to my colleague the late Vladimir Markotic to get his opinion. He agreed with my conclusions. I don't know if Guy Phillips agreed though; I have not heard from him since.

As with footprints, there have been a number of cases of mistaken identity involving sightings of the animal itself. I remember one case in recent years where a fellow named Mike Hammer (yes, this is his real name) phoned me up to report he saw what he thought was a sasquatch in the Morley Indian Reserve, west of Cochrane, Alberta, as he was driving back to Calgary at 2 A.M. on the morning of June 6, 1989.

"It was just standing on the side of the road! Up on the side of a hill there. I only saw it for a few seconds as I drove by." Mike was adamant he saw a sasquatch and he suggested I go out to the spot with him to see if any tracks were still around. On June 8, I did go to the spot with him, and what we found was a large lone rock, which has been on this spot as long as anybody remembers. The rock could in my opinion resemble a large creature in the dark, especially when someone drives by at high speeds and only gets a fleeting glimpse of it. Mike was embarrassed at having wasted my time and apologized for having dragged me out there. I told him no apology was necessary. I was grateful he at least reported what he thought he saw, particularly since most people who may have really seen a sasquatch will not report their encounter to anybody.

I recall another phone call from a woman, whose name I cannot remember now, who told me about a sighting she had made while she and her mother were driving out of Harrison Hot Springs on their way home to Vancouver, at 3:30 A.M. in August, 1985. She saw at the side of the road a sasquatch that appeared to be about to throw something at their car. She was feeling sleepy when the two women started their trip home and she was sort of dozing against the passenger's window when she saw the animal. She called out a warning to her mother who was driving. Her mother was concerned about her daughter's fright, but she did not stop. Later when the daughter phoned me to tell me of this incident, I asked if this actually happened inside the town limits. She told me yes.

There have been previous reports of sightings inside Harrison Hot Springs that I consider to be true, but not this one. As she was talking to me, I recalled the first time I ever entered Harrison Hot Springs early one morning in 1981. Being the early morning, most of the resident's outdoor lights are turned off, including the lights that illuminate the sign at the entrance to Bigfoot Campgrounds. Beside the sign stands a large wooden carving of a sasquatch holding a large boulder over its head. When the lights are turned off, the sign is hard to see. I also nearly brought my truck to a stop the first time I glimpsed this statue standing there, looking at as though it was about to throw this rock at my truck. I think this woman was fooled by this same statue. Of course, she didn't stop to take a closer look either.

I also have a personal experience with a case of mistaken identity, which happened during the summer of 1987. I was driving west along Route 99, between Lillooet and Mount Currie, B.C., with my ex-wife. This was the first and only time she came along with me on a research trip.

Route 99 today is a paved highway. In 1987, it was still a dirt forestry road. The sun was low in the sky and it was just getting dark. We were driving along slowly looking for any tracks that would indicate a sasquatch had crossed the road. Suddenly she cried out, "Stop!" I asked her what she had seen, and she replied that she thought she had seen a large creature in a clearing walk from one tree stump to the next. There was a river between us and the hillside where she thought she spotted this thing, and I was determined to make sure she actually had seen something before I was going to get wet.

I put the car in reverse and backed up some distance. I then started to move forward again, instructing her to see if she saw it again. She did. In the clearing it appeared that a forest fire had gone through this area some years previously and it left several large blackened stumps on the hillside. The creature turned out to be a stump slightly shorter than the rest, and a little further away, with a bit of a nob on top which she took to be the creature's head. It was the fact we were moving that made it appear like some creature walking from one stump to another. I personally was disappointed that it wasn't a sasquatch, but at least I didn't have to get wet.

My companion today is much more interested in my research into the sasquatch phenomenon. Barbara and I were camping at Burnt Timber Creek, in the Bow Crow forest of Alberta during the summer of 1995. We were spending some time together walking along the creek when she pointed out an old fallen log that lay at the end of the tree line on a large hill, across the creek to our right. "Doesn't that look like a sasquatch sitting up there," she asked. Yes, I had to agree with her, it did. However, it actually was the root system of a fallen tree. But to the untrained eye it could appear to be a sasquatch sitting there with both its legs bent at the knee, sitting like a man watching the valley below. Even through the binoculars it looked like some creature sitting up. We both talked about what would happen if somebody was driving through this area and had just glanced up and spotted this fallen tree. Would they think it was a sasquatch?

Another man phoned me to describe how he had mistaken a bear for sasquatch while he was out hiking near Radium Hot Springs, B.C., in May, 1991. When he came across my first book *The Sasquatch in Alberta* in a bookstore in Banff, he decided to tell me about his mistake, hoping I would get a kick out of it. He was right.

He was on a hiking trail when he saw what he at first thought was a sasquatch leaning against a tree. He thought to himself, "Wow, I've really seen one of those things." As he watched, the sasquatch went back down on four legs and ran into the trees. It turned out to be a black bear getting a good back scratch against the tree. The gent laughed to himself and went back to his car in case the bear was still around.

Certainly, mistaken identity, can explain away many reported sightings of sasquatchlike creatures. It even may be the explanation for the Wayne Oliver video of May, 1997, which I discussed in chapter three. It may have been a large man he and his companion videotaped that day. Though, he personally doesn't think this could be so. Mistaken identity or visual misinterpretation, the term I like to use, certainly cannot be used to explain all sightings. There are too many cases where a witness was able to watch the creature for sometime and also observe the creature walking, running, feeding, drinking, grasping, leaping and/or watching the witness, as it seems to continually check out its surroundings in a nervous manner. However, science needs evidence that can be laid out on a table and physically examined. In the case of the sasquatch, this means a body, piece of a body or skeletal remains need to be presented. The evidence we have so far, if you weed out the cases of mistaken identity and hoaxes, is very convincing, though circumstantial, in favor of the creature's existence. If some day it is proven this creature does not exist or never has existed, I have no problem with that. I was simply wrong.

The Alberta Scene

When I moved out to the Canadian west in 1979 from Ontario, I was serving in the Canadian Army, with the First Battalion, Princess Patricia's Canadian Light Infantry (PPCLI). Joining the army when I was eighteen was my ticket west, having had the sasquatch mystery on my mind from early on. When I had completed Battle School, I requested a posting to the Third Battalion, which was then based in Esquimalt, B.C. However, the army had other ideas.

I ended up in the First Battalion, based in Calgary, Alberta. I was disappointed at first, since my goal was to live in B.C., the traditional range of Canada's hairy giants. However, I thought I would make the best of it since Calgary is only a hop, skip and a jump from the eastern part of B.C. I knew there had been a few reports in the province of Alberta, having read as a boy about a few in John Green's earlier booklets, as well as his major work *Sasquatch, The Apes Among Us*. Nevertheless, I was at first just biding my time, and doing my duty of course, until I could one day arrange a transfer. As soon as I realized that Alberta and British Columbia both shared the beautiful Rocky Mountains, I thought to myself, "Well, there is no wall between the two provinces. If sasquatch has been reported in eastern British Columbia, it must have been seen in western Alberta as well."

Aside from the few stories in Green's books in which the witnesses had contacted him in order to tell of what they had encountered, nobody to that point in time, 1979, with the exception of my late friend Vladimir Markotic, had made any attempt to catalog research and follow up on reports from the Alberta side of the Rockies. Vladimir never advertised for reports from Alberta, but he had a general interest in the subject.

When I first met him in 1980 and told him of my intention to track down Alberta reports, he took me under his wing, so to speak, and encouraged me to do so. We stayed good friends until his death in 1994. When he died, it was like I lost a favorite uncle; I still think of him often. With his encouragement, since I was still mostly unknown to sasquatch researchers in 1980, I figured the best way to let witnesses know I was around, was to advertise in Calgary newspapers. I asked witnesses to contact me if they wished to tell anybody about what they thought they saw. The result was almost overwhelming. I was answering my phone at first almost nightly, talking to hoaxers, the lunatic fringe, reporters, as well as people who really believed they had seen the creature. Most of the calls concerned sightings in B.C. However, a large amount came from the wild areas of Alberta. The result of all this was my first book, *The Sasquatch in Alberta*, which was published by the now-disbanded Western Publishers in 1990.

At the same time, I would travel to B.C. at every opportunity to do research there. During the 1980s I also made many trips to the Pacific Northwest of the United States as well, Washington, Oregon and Northern California. I haven't been south of the border since 1989, when I gave a paper at the conference in Pullman, Washington, sponsored by the International Society of Cryptozoology (ISC), due to the lowly Canadian dollar. Today when I get a report from the Pacific Northwest of the U.S., I simply pass the information on to the closest American researcher I know, and let him/her follow up on it. He or she in turn informs me when they hear of something in eastern B.C. or Alberta.

As I said, it was almost like opening flood gates when I started to advertise for Alberta reports. It seems many Albertans thought of the sasquatch as a B.C. phenomenon, and few knew anything about Alberta reports. Well, that idea soon vanished. I will begin this chapter with one of the more recent reports I have received.

A gentleman from Cochrane, Alberta, contacted me to tell about a strange animal he saw while he was camping off the 940 Trunk Road, about four miles south of Livingstone Falls Recreation Area early on the morning of August 4, 1996. Bruce Adams (not his real name) was camping in the area waiting for a large dirt bike rally that was to begin that weekend. Early in the morning of August 4, he made himself a cup of coffee and took his dog for a walk. During the walk he saw an animal that left him questioning his own powers of observation. My full interview with Bruce Adams follows.

Q: "Please state your full name."

Real name withheld.

Q: "Where did this incident occur?"

A: "It was four miles south of Livingstone River or Livingstone Campgrounds on Highway 940."

Author's note: The actual name is Livingstone Falls Recreation Area, though camp sights are available. Also many people refer to the 940 as a highway, seeing as it runs from Highway 3 in the south of the province to Hinton, west of Edmonton. Most of its length today remains an unpaved forestry road.

Q: "What date did this take place?"

A: "August 3rd, it was the Sunday of the long weekend in August, 1996."

Author's note: He had his date wrong. August 4th, was the Sunday.

Q: "Was it at night or day?"

A: "It was morning, about nine o'clock."

Q: "Describe the area for me in which it took place."

A: "It's a wooded area, where the camp was. And where the road runs through, it's an open clearing, just low bush and tall grasses, surrounded by mostly pine and spruce."

Q: "What distance would you estimate you were from this thing when you saw it?"

A: "Probably about 1,200 feet, maybe a little farther."

Q: "Heading south on the 940 Trunk Road, did this sighting occur on the left-hand side of the road or the right?"

A: "On the right-hand side."

Q: "The right?"

A: "It would be the west side."

Q: "Were you on the right-hand side when you saw it?"

A: "No. I was on the east side."

Q: "So you were down more toward the river?"

A: "Yeah. I was down beside the river and I was looking across 940 toward the far side of the clearing on the opposite side of the road."

Q: "What was your first impression?"

A: "The situation really, your mouth drops, you know, it's just to see another living being, ah, like anything in the wild. If I'd seen a bear or a moose I would have, you know, just stopped, enthused,

132

to be able to see it and my German shepherd was with me too, and he was just focused on this thing as well."

Q: "What was the dog's reaction?"

A: "Just full attention—ears forward and pointing straight toward it."

Q: "Didn't bark or anything?"

A: "No."

Q: "How did you react when you saw it?"

A: "Well, I just wanted to see more. He was moving. I sort of just froze and was watching."

Q: "What was it doing?"

A: "It was running...seemed to be moving north to south at the far edge of the clearing near the trees. He didn't seem to be putting a lot of effort in, but he was covering the ground very quickly."

Q: "Did it stand and move on two legs?"

A: "It was on two legs, yes."

Q: "Did it ever go down on all fours?"

A: "No."

Q: "Was it hairy?"

A: "I would say yes. It was all black and it certainly wasn't all...clean cut. I would have to say yes, definitely. It was covered with hair. The color was all black."

Q: "How tall would you estimate it to have been?"

A: "I'd say better than six feet. I'm trying to compare it to...the next day there were lots of people standing around. I'm trying to compare it to them. I would say six feet or more."

Q: "How heavy do you think it was?"

A: "Hard to tell, it wasn't slim. I would say it was between two to three hundred pounds."

Q: "Could you see any facial features?"

A: "I couldn't see any detail on the face at all...I could tell there was a head there, but no detail at all. No real definition of a neck or detail of the face."

Q: "Could you describe the arms?"

A: "I didn't...no visible arms either, at least there was no arm movement. When I run, my arms move. It was moving and I couldn't see the legs either because of the bushes and the grass. So basically what I saw was something moving swiftly across the ground."

Q: "So if it had arms, what you are telling me is that when it was moving, it kept them straight down at its sides."

A: "Yes."

Q: "Could you tell if it was male or female?"

A: "No."

Q: "How long do you think you saw it for?"

A: "Probably fifteen seconds."

Q: "Did it see you?"

A: "No. He was focused on something ahead of him or something in the bushes, at least it appeared that way to me."

Q: "Did it make any noise?"

A: "Yes."

Q: "Could you describe it for me?"

A: "Well I could try to replicate it."

Q: "Sure, go ahead."

A: "It was the noise we heard first of all, and that's what got the dog's focus as well. Sort of a..."

Author's note: He made a noise we both agreed later was something like the wail of a bear cub. He continues.

"That noise never stopped, as long as I saw him running he was making that noise."

Q: "Kind of like a bear cub in distress will sometimes make?"

A: "Yeah."

Q: "Sort of howling?"

A: "More of a calling. The impression I would put to this is that if it had been a bear cub and he had lost sight of its mother and was trying to smell and see her and was moving quickly to catch up to her or where he thought she was. Sort of describes it. It didn't appear to be running from anything."

Q: "You saw this thing from profile?"

A: "Yes."

Q: "Did it seem to have a snout?"

A: "Nothing that I could see. I couldn't see any detail at all."

Q: "Did you smell anything before, during or after?"

A: "No."

Q: "Are there any other physical characteristics of this animal that stand out in your mind?"

A: "No. No, I guess that the only comment would be that it seemed to lack any specific body shape. I would almost describe it as a cucumber with hair. Something without very much definition that I could see at all."

Q: "Did you check for footprints?"

A: "No. I didn't go over there."

Q: "Did you report what you saw?"

A: "I called you."

Q: "Nobody else?"

A: "I told my mother what I saw and she said it was most likely a bear. It got to the point where I was thinking bear, but the tallness of it and the fact it was on two legs...a moose from the front running sideways? I had all kinds of ideas as to what it was. A bear didn't come to mind right away."

Q: "In you own words describe what happened."

A: "We were camped below this meadow so I climbed this steep bank. When I got to the top of the bank, I was level with the meadow. I took a few steps forward and then I heard the noise it was making. I stopped in my tracks and the dog stopped and we both focused forward. I looked up and I could see him against the trees on the far side of the meadow and he was traveling from where I was standing, right to left (north to south). And while he was running I was watching and listening. I was hoping he would stay out in the meadow so I could have a longer look or even come closer. But after about fifteen seconds, he disappeared into the trees. I watched for awhile but never saw any other signs."

Q: "Were you walking your dog?"

A: "Yeah. I had my coffee in my hand and was just walking around. The reason I was down there was because of the cross-country motorcycle race. It was two to three hours before the race would start and I was expecting my brother down there so I was just going for a walk along the edge of the campsite and see if I could find him."

Q: "Did you see or hear it first?"

A: "I heard it first and that's what drew my attention to it."

Q: "Do you think it was making this noise because it spotted you?"

A: "No. I don't think he looked over at me at all. He didn't appear to be running from anything. He was in full sight of quite a few trailers. There were quite a few people there for the race so, no, I don't think...I don't really think that he was aware I was there. I have seen black bears running from me before, when I was on a motorbike, and they're down on all fours going like crazy."

Q: "Did this thing seem fluid upright or awkward upright?"

A: "It seemed very fluid. There was no awkwardness about it. It just seemed to be moving along like a person on a bike, behind a hedge so you couldn't see the bike behind a hedge so you couldn't see the bike. Just the body moving across."

Q: "So what you saw was something in about waist-deep bushes, so you saw it from the waist up in profile?"

A: "I would, I would suggest that the bushes were...just over knee high and I saw more than that. I saw below his waist, but it seemed with the amount of hair, and the way he was running...I could tell there were legs moving, but I couldn't see the full legs."

Q: "You didn't see anything in profile, like female breasts or anything like that?"

A: "No."

Q: "It never went down on all fours?"

A: "No."

Q: "In your opinion...you were talking to me earlier and it seemed like you were trying to convince yourself that you saw a bear. Or do you think you saw something else?"

A: "It could have been a bear, but I have doubts that it was a bear...I'm trying to keep an open mind as to what it might have been. I intend to talk to some bear experts and find out more about the behavior of bears for one thing. I can't think of any other animal it might have been."

Q: "Do you think it could have been a man?"

A: "No. It was moving too quickly and too effortlessly to be a man. The size is deceptive. A man in a bear costume? Ha. Ha. It would be pretty awkward to run that fast. No, I didn't recognize the motions of a man."

Q: "Are you bothered by what you saw?"

A: "No. I accept it. I saw something I just can't figure out what it is. I wasn't scared or...I was sort of awe struck. Whatever it was, I was elated to have witnessed it."

Interesting that Bruce Adams never states during this whole interview, "I saw a sasquatch." He does think that due to certain physical abilities of the creature it was not a human. He still won't dismiss a bear entirely though because it was big and was covered in black hair. Before this happened, he did not believe in the existence of the sasquatch. He's still not sure if he believes in them now. A rare opinion from a person who might have seen one. He only contacted me because he couldn't find an answer himself as to what it was he saw. However, I'm prepared to dismiss a bear on the same grounds as he dismisses that it was a person—bears do not run at fast speeds up on their hind legs.

The creature he saw apparently could. He also thought it moved too fast to be a man. I went out to the area one week after this incident was reported to have taken place. I found no evidence, tracks, hairs or anything else, to confirm that a sasquatch was here one week earlier. I really didn't expect to find anything though. If there were footprints, I'm sure that the cross-country race that was held there later that day would have destroyed them. The race, by the way, is an annual event held each August in this area. I'm sure any sasquatch, bear, moose or any other animal in the vicinity would move far away when the racers with all their bikes, campers and noise come into the area. Bruce says he's not bothered by what he saw and I believe him. However, I'm sure he wanted answers, that's why he contacted me in the first place. Answers I couldn't give him due to the fact I don't know if the sasquatch exists. I believe they do, but I don't know for sure if they do. If they do, I think he saw one.

As I said earlier there were reports of sasquatchlike creatures in Alberta before I came along in 1979. There was a series of events that occurred in the fall of 1972 and October, 1973, near the small Alberta community of Seven Persons, which is located a few miles southeast of Medicine Hat. Seven Persons is a very small settlement with a few grain elevators, a couple of stores, a gas station and not much else. The population consists of hard-working farmers and ranchers in and

The clearing where Bruce Adams (not his real name) saw a strange animal, which fits the description of a sasquatch, near Livingstone Falls Recreation Area, Alberta, August 4, 1996.

Photo: T. Steenburg, 1996.

around Seven Persons who make their living off the land and are the type of people who made Alberta the proud, conservative-minded province that exists today. It is also a prairie community, several hours drive from the wild forests of the Rocky Mountains.

If these events were to occur today, my reaction would be one of skepticism due to the fact sasquatch country is so far away. However, there is some tree coverage along the South Saskatchewan River, as well as Cypress Hills Provincial Park, which is located about an hour's drive southeast. Still, it is not an area where one would expect reports of sasquatch. At least not true reports. Nevertheless, for a period of about one year from fall, 1972, to fall, 1973, there were a series of reports that turned the local population upside down. John Green received reports from two sources in the area at the time. One was a local rancher, who for a time found himself being the local sasquatch hunter. The other was from a fifteen-year-old teenager, who also had an interest in what was going on.

Leonard Edvardson, a local rancher who started looking into some of the reports and also photographed some tracks, went on local television (CHAT-TV) and suggested maybe a sasquatch was in the area. As a result, people started contacting him to tell about what they had seen. A couple of searches were organized, but turned up nothing. The first of several reports over the next year or so happened during the fall of 1972. Two boys out walking at a place called Police Point said they saw a large, upright-walking creature, not too far away from them. Nobody at the time took much notice of the boys' story. It was nearly a year later when another report came in. In early December, 1973, at about 3 A.M., a man reported a similar creature inside the village of Seven Persons, which he could see clearly due to bright moonlight. Tracks were found the next day on the ice of a nearby creek. The tracks were reported to be about fifteen inches long and seven inches wide. The step distance between each print was reported to be about six feet. The man who saw the creature reported that it stood between seven and eight feet tall. He also said it looked as if the creature was missing one forearm or else it had one arm doubled up.

Later two men who were ice fishing at the Murray Dam reported finding tracks in the snow, which they said had a step of about six feet. The men were concerned whether or not it was safe to continue fishing there.

Another farmer came forward to report something had frightened his cattle, causing them to break out of their corral, four nights before the sighting in Seven Persons. Now people were becoming worried, and reports of large apemen in the area and stampeding cattle was something that could no longer be ignored. Two searches of the area took place on December 13th. Aircraft were also used. However, nothing was found. Local television and radio became involved. A zoologist from Medicine Hat College, Professor Gallaway, said on CHAT-TV that as far as he was concerned the whole thing was a lot of nonsense and photos of the tracks showed common animal prints that had melted out due to warm weather. After the professor gave his opinion excitement died down. As far as most locals were concerned, some kids must be having everybody on with a good prank. Mr. Edvardson did receive two letters after though from a man who would not reveal his identity. He simply signed the letters "An Observer."

The enclosed sketch is of something or someone I saw at twilight in Medicine Hat, in Kin-kouly. At first I thought it was a bear yet knowing this could not be I watched for about 4 minutes the figure which seemed to carry something in its arms. After this it vanished in the small creek and bushes and I could not see it anymore. I estimated it to be at about 6 or 7 feet tall, but going with an awkward and bent gait. I was alone and there was nobody else around to show what I had seen. Maybe it has been some kind of Halloween joke, but I thought no more about it until I read your letter in the news. When I came home, I made several sketches, but didn't show them to anybody. If you are interested, I could copy some of my other sketches. I am not giving my name at this time. I do not want to get any phone calls about it.
—An Observer

Right: Sketch sent to Leonard Edvardson by witness who identified himself as "An Observer" of a creature he saw near Seven Persons in December, 1973. The creature appeared to be carrying something in its arms.

Second sketch sent to Leonard Edvardson by the Observer of the creature he saw in December, 1973. Again the creature seems to be carrying something in its arms—a female with a baby perhaps?

140

Mr. Edvardson received a second letter from the Observer a few days later; it gave a little more detail of the creature that was seen. Two more sketches were included with the letter.

Dear L. Edvarson:

I copied two more sketches of the ones I did that day, or better said, the evening I saw what I tried to bring down on paper. The "Thing" was about 250 ft. Far away and going sideways at a very odd gait. It seemed to have fur and no clothes from reddish brownish dark color. It was going slow, not hurrying. Carried something in its arms like a bundle. I still think its being an early Halloween "spook," but then, why would the figure have entered the cold creek? Wouldn't that have been carried too far by some jokester? I saw no one else watching and as I said before it was dusky already. The head I did draw certainly not so clear, but I tried to imagine what it looked like but the profile is right in its outer lines. I will not call you as I don't think there's anything to it that should be made public, —I went there again, several days later in bright daylight and I could see nothing where it had walked. But the creek was deep on this place.

—An Observer.

P.S. The legs looked short and thick.

The two letters and sketches from the person who called himself the Observer, marked the end of reports of sasquatchlike creature around Seven Persons. The two letters had apparently been written on an old typewriter in need of a new ribbon. I've reproduced them here exactly as they were written. I've made no attempt to correct spelling errors or grammar, for I feel that these, like the interviews, give an indication to the reader of the personalities involved. No other reports that I know of were made. It has been twenty-five years since these events occurred and, as I've said already, as far as I know there have not been any sightings of creatures near Seven Persons since. Part of me feels that this story should be in either chapter seven or eight. However, they do have a ring of truth to them, and I wonder why such a creature would be in this area. They did receive a lot of attention from local media at the time, which is apparently why the Observer did not want to reveal his or her identity.

In chapter three I wrote about roadside sightings in B.C. Alberta, too, has had its share of this type of encounter. However, I will not devote a whole chapter to them. John Green received a letter from a twelve-year-old girl from Edmonton who, with her parents, reported seeing more than one creature cross Highway 11 (David Thompson Highway), which runs between the town of Rocky Mountain House to Banff National Park. Along this highway lies the small community of Nordegg. Also, the Big Horn Dam and Abraham Lake are along this highway. This area seems to be where most Alberta reports come from. In this case five or six creatures were reported crossing the road in front of their family car. Since the girl was only twelve at the time I will assume anonymity was requested. The letter was written on December 27, 1976.

Dear Mr. Green,

I have heard that you are doing research on the Sasquatch. I hope this letter is of some use to you. This incident happened in August last summer (August 1976). We were on our holidays and we had just gone through Nordegg Alta (Alberta). We were driving in our car pulling our trailer at about 60 mph. Mom and Dad were talking in the front seat. I was reading a book in the back seat when all of sudden Mom and Dad stopped talking. I thought this was strange so I looked up and said, "Anything wrong?" Dad and Mom were staring straight ahead. Finally Dad said, "Shelly, look!" I looked up and then I saw them. About half a mile away there were five or six things (sasquatches) walking across the road, upright. They crossed the road in single file, from tallest to smallest. They were all black and had long hair. They seemed to have been very tall from four to six feet or more in height. One thing that was quite strange though, we were the only car on the road and the Sasquatches crossed the road without turning their heads from side to side. It was as if they knew it was safe to cross the road. One thing is too bad though. There was a pair of binoculars and a camera beside me, but at the time I had completely forgotten. I hope this letter helps you. By the way my name is Shelley Snow (not her real name), and I am twelve years old. This incident happened in August and Mom and Dad saw these creatures with me.

Yours truly,

Shelly Snow

Well as you can see Shelly seemed to be a very intelligent little girl in 1976. Today she would be a woman of thirty-four years of age. I have tried to locate her now without success. It is too bad that she never thought to take a picture of the creatures. A photograph of six sasquatch crossing a highway would have been very interesting indeed. However she shouldn't have felt too bad, many adults before and since, with the exception of the late Roger Patterson and perhaps one or two others, have neglected to take pictures.

One day I received a phone call from a man named Darren Romeo who, with a friend named Barry Barbetts, saw a strange animal while they were walking on a cutline a few miles northwest of Rocky Mountain House in late April, 1987. The two men had been enjoying the splendid scenery when they noticed something on the cutline not too far ahead of them. The creature stood there for about three seconds, then it ran into the trees on the right-hand side (east). The creature was upright on two legs the whole time it was in view.

"I never, in all my years of hunting, seen anything like it!" said Darren. Both men told me they thought the creature stood between seven and eight feet tall, and it was covered with dark brown hair. It had happened so fast (about three seconds), neither man said anything to the other for about five minutes after the creature disappeared into the forest. Darren later said that the two of them must have looked fairly silly just standing there staring at the spot where the creature had been standing. The men then decided to return to their truck and leave the area.

George Harris from Squamish, B.C., then living in Nordegg, Alberta, reported that in September, 1968, he photographed several large humanlike footprints that were found by some of the local Native residents. The alleged tracks were found at the spot where the Cline River flows into the North Saskatchewan River. The prints were said to be seventeen inches as well as thirteen inches long. It appeared to be the tracks of two individuals. However, no copies of these photos have been added to my files and I have not examined them.

Not long after these tracks were found, reports started to come in from Natives who had decided to live off the land as their ancestors had done. One young man, Alec Shortneck, reported that while he was chopping wood, he looked and was surprised to see the head and shoulders of a large creature watching him doing his work.

"I didn't know what to do," he was quoted. "So I just carried on with my chopping. When I looked up again, the thing had left."

Edith Yellowbird, a sixteen-year-old young woman, reported that she and three other women saw four creatures up on a mountainside near Windy Point.

"I think they had caught something. Two were bending down and the other two were just walking about near by. They were as tall as good sized spruce trees on the mountain side on which they were standing." Edith was with three middle-aged women who confirmed her story.

More reports came in and many people of Nordegg were convinced that a group of creatures was living in the forests around the town. All this lead up to the now-famous Big Horn Dam incident.

I have written extensively about this episode in my two other books. However, new information has come my way which throws a shadow of doubt on the whole thing.

In 1969 the dam was still under construction. On August 23, 1969, five workmen working on what would become the water pumping station for the dam reported watching a fifteen-foot creature watching them work from a high cliff about a mile away. Harly and Stan Peterson, from Condor, as well as Floyd Engel of Eckville, Guy L'Hereux of Rocky Mountain House and Dale Boddy of Ponoka all said that they watched as the creature walked, sat down, got up and continued to walk all along the top of this cliff. The men stopped working and gathered together to watch it.

Harly said, "It looked like an incredibly large man, enormous, head slightly bent forward and very hefty." Dale Boddy reported that it was too tall and its legs were too thin for a bear. As well, the speed it was moving at approximate strides of six feet in length. (Author's note: It is possible he may have been talking about the steps rather than stride. However, I'm not sure of this.)

None of the men had binoculars or a camera. When the creature turned away from the edge of the cliff and disappeared into some trees, it had been in sight for about forty-five minutes. Later, two of the men went up to the spot where it was and the men who stayed

The high ridge at the Big Horn Dam, where five men reported seeing a fifteen-foot creature watching them as they worked, on August 23, 1969. This incident was for years considered Alberta's most convincing case supporting the existence of the sasquatch.

below were amazed that their companions at the same spot were less than half the height of the creature. They all later agreed that the thing was about fifteen feet tall. The press got a hold of the story and this incident received a lot of media attention. Both John Green and Rene Dahinden came to the Big Horn Dam to investigate. Rene even took a job there in order to stay for a while and did not leave until the Bossburg Cripple tracks were found in northeastern Washington a few months later. This incident for years was considered Alberta's most convincing sasquatch report. It remained so until the Crandell Campground incident of May, 1988, which I will go over later in this book. It was reported at the time that many of the local Natives were against the dam's construction, though many of them were happy to have the jobs at the dam's site. As I wrote earlier, I now have some new information that I was not aware of when I wrote my other two books.

In March, 1994, I was told that some of the local Natives have a ceremonial sasquatch costume that is used during ceremonies in this area. The man who wears this costume apparently has to be trained to walk about on stilts. I was told that this costume was in use in 1969. I have to wonder if this costume was used at the dam site that day to try to scare workers away. I have spent three years now trying see this costume, but the local Natives become tight-lipped when asked about it so I cannot say for sure whether or not it does exist. But I have been told that the wearer cannot move about very fluently. The creature seen by the five men apparently could, as it was observed walking the whole length of the high cliff. So maybe it was not used that day by Natives trying to stop the dam's construction. However, I feel this possibility should be looked into further, which is what I intend to do.

In 1972 a thirteen-year-old girl, named Debra, had a frightening encounter with a creature that fits the description of a sasquatch. Her family was camping just off a forestry road, just west of Nordegg and north of the construction site of the Big Horn Dam. While her father was busy cleaning some fish he had caught, she wandered from the family camp site on her own little nature walk. After a short time she came across some ripe berries, and bent down to pick and eat some. As she picked berries, her nostrils were invaded by a very unpleasant stench. She later described it as "not as bitter as a skunk smell, but muskier than that of a dog in heat."She then had an overwhelming feeling that she was being watched. She whirled around, half squatting, half standing, and found herself facing a creature like nothing she had seen before. It was no more than fifty feet away and in plain view standing next to a thick line of bushes. It stood about eight feet tall, upright like a human. Its arms hung down past his knees, its shoulders were about three feet across; it was covered with long, matted, dirty brown and gray hair. The crea-ture's nose jutted out from its face like a flattened ball of putty. However, it was the creature's eyes that caught her attention. They were slate black, scrutinizing, even scary.

"That's no moose!"was her first thought. She knew it wasn't a bear, so what was it? She then stood up; her eyes met its. It was standing up just like a person would but as far as she was concerned it was an animal. It didn't do anything. It just stood there staring right back at her. After what seemed like forever to her, the animal let out a deep, throaty moan. Nearing a state of shock, she turned and ran, fighting her way through the trees, trying to remember the

146

way she came. A wave of nausea hit, she stopped to vomit. When she did find her way back to the family's camp site she burst through the trees smelling of vomit and had tears rolling down her cheeks. Her face was pale, she shook all over and was barely able to walk. When she found her brother, she tried to explain what had happened. Her mother overheard the story and was convinced her daughter had been frightened by a wild animal of some kind, possibly a bear. She listened, but did not believe that Debra had seen what was described. Debra slept with her parents that night. Later she was scolded for holding fast the story of "the big hairy man in the woods."

Apparently Debra at this stage of her life had never heard of the sasquatch, so she had no idea what it was she saw. It was about one year later, as part of a school assignment, she learned that many people had seen something very similar to what she had. She did a major school project on the subject doing her presentation on what she had read in books, never letting on that she had seen one herself. Her teacher was very impressed with her work and told her, "Well done Debra, it's like you've seen one yourself." She still didn't mention her encounter a year earlier.

Growing up she would mention it from time to time to friends at parties when the conversation would turn to the unknown, she would make the simple statement, "I've seen a sasquatch." Her story was told less and less over the years, but she could not put it out of her mind. Eventually the experience invaded her dreams at night causing nightmares. Finally she came across my ad in the Calgary press. She did not want to phone at first, it was some time before she dialed my number.

Finally on January 25, 1988, she phoned and told me of her experience, and we arranged to meet at her home the following evening. Her husband as it turned out was a soldier with my old regiment (PPCLI) and he said to me, "You know, I don't believe in any of this, but my wife has stuck to this story and swears that it did happen." When I met Debra Malone my impression of her was that she was a charming, intelligent woman who seemed relieved to finally talk to somebody who wouldn't judge her before they heard the whole story. Following is a transcript of the interview.

Q: "Where did this incident occur?"

A: "It was near Nordegg, near a big lake. I believe it is called Abraham Lake. It was on a forestry road...I would say about twenty miles into the bush."

Q: "North of Highway 11, or south?"

A: "I think more west."

Q: "What date did this take place?"

A: "Near the end of June, 1972."

Q: "Was it at night or day?"

A: "Day."

Q: "What time of day?"

A: "It was about 3 o'clock P.M."

Q: "Describe the area in which this took place."

A: "It was fairly bushed, there weren't any gigantic trees or anything around—a lot of low-lying shrub. There were a few tall trees, pine, fairly green, you know, nice."

Q: "What distance would you estimate you were from the animal when you saw it?"

A: "I have thought about this time and time again. I would say I was not more than fifty to fifty-five feet away."

Q: "What was your first reaction?"

A: "I was stunned."

Q: "What was it doing?"

A: "Standing there."

Q: "Did it stand and walk on two legs?"

A: "Yes, it did."

Q: "Did you ever see it go down on all fours?"

A: "Never."

Q: "Was it covered in hair?

A: "Yes."

Q: "What color was it?"

A: "It was matty...matted brown, with some gray in spots. Different shades of brown."

Q: "How tall would you estimate this creature to have been?"

A: "Over seven feet."

Q: "What would you estimate its weight to have been?"

A: "Oh, I have no idea. It was big, heavy."

Q: "Did you see any facial features?"

A: "Yeah, I did, eyes, nose. The eyes were most prominent."

Q: "Could you describe them?"

A: "The eyes were set fairly back in the head...I don't know, they were dark, very deep set. The nose was kind of jutting out a little bit, and came down in kind in kind of a slope. That's really the best description I can give you."

Q: "Could you describe its arms?"

A: "Very long, hung down almost to the knees."

Q: "Were they very big?"

A: "Yep. This thing had gigantic shoulders. I would say at least three feet across."

Q: "Could you see if it was male or female?"

A: "No, I couldn't. It didn't have breasts or anything. I know that much."

Q: "How long did you see this creature for?"

A: "Well, it seemed like an eternity. I would say I stood there and stared at this thing for about seven to ten minutes."

Q: "And it stood there and stared at you?"

A: "Yep, it stood there and stared right back at me."

Q: "Did it ever make any noise?"

A: "It made a low moan. It sounded like in the back of its throat."

Q: "What was its reaction when it saw you?"

A: "I think it felt the same as I did. It was amazed that I was there, and I was amazed that it was there. We were both trying to figure each other out. That's my impression."

Q: "So you were both just standing there looking at each other?"

A: "Yes, exactly."

Q: "Did you smell anything?"

A: "Yeah, I did. It was musky. The closest thing I can tell you what it smelled like was a dog in heat. That's the only thing that I can think of."

Q: "After it had moved off did you check for any footprints?"

A: "No, I didn't."

Q: "Did you report this to the police?"

A: "No."

Q: "Did you report it to anyone else?"

A: "Just my brother."

Q: "In your own words, describe what happened."

A: "Well, weren't doing anything that afternoon. My father was cleaning a bunch of fish he caught, so I decided to take a nature walk through the trees. I guess I strayed some distance away from where we were tented out. I remember walking and walking and walking. I came across some low-lying plants. I bent down and started picking them when I heard some bushes rattle. I turned around and I saw this creature standing there. I rose to my feet immediately and looked bold-faced right into this creature's face. It stood there and looked back at me, for the longest time we just stood there and stared at each other. It made this low noise in the back of its throat and that was like instant reaction. I just took off and I ran as fast and as far as I could. After I was away for a little, while I remember getting sick. I was so scared I was sick! After being sick I ran the rest of the way back to our campsite."

Q: "How old were you when this happened?"

A: "Thirteen."

Q: "You were on a camping trip with your family?"

A: "Yes."

Q: "And you wandered off on your own?"

A: "Yeah. I just wandered off on my own little nature thing."

Q: "So you were the only one to see this thing?"

A: "Yes, unfortunately."

After the interview Debra gave me a faded pen sketch, she made when she and her family returned from that camping trip in 1972.

In the years that followed my interview with her, it was like she had a weight lifted from her shoulders and agreed a couple of times to appear on radio talk shows that wanted me to provide some witnesses to tell of their personal encounters with the creature. However, all these fell through when time constraints on the programs forced the radio interviewer to drop the witness segment. When I was asked to appear on the CBC television show *On the Road Again*, hosted by Wayne Rostad, at first they wanted to add some eyewitness testimony from the witnesses themselves on the segment. Debra had agreed to appear after some coaxing from me. Again, however, time constraints forced the show to instead talk to some of the local people in the Water Valley Saloon, which is the town I was living in at the time. I have now lost contact with her. I assume she now lives in Edmonton, since her husband is a soldier and the base here has been shut down and all personnel have been moved to a new base up there.

The only thing that bothered me about her story was the description of the creature's smell. She described it as similar to a dog in heat. I don't know anything about dogs, but I've talked to people who do and as far as they are concerned a female dog in heat does not give off any kind of smell, at least any kind which humans can detect. So I'm a little confused over that description. However, as to her integrity among her family and friends, I have no doubt. They may not believe she saw a sasquatch back in 1972, but they all agree she must have seen something.

The next two interviews were from hunters. Both incidents again occurred between Rocky Mountain House and Nordegg. Hunters for the most part are individuals who are clear headed and not likely to mistake some common animal for a sasquatch. Hunting regulations in Alberta are stringent and enforced. Hunters have to make sure they only shoot an animal for which they are licensed. If they shoot something else, they face hefty fines and possible jail time. The two following cases are not only interesting because deer hunters reported seeing a sasquatch, but in both cases neither man was alone.

The first happened in October, 1982, in the wilderness area near the Nordegg River. Miles Jones (not his real name) was out hunting and fishing with a friend, Wayne Farson (not his real name). The two men hadn't had much luck finding the deer they were after, so when they reached the spot where Owl Creek flows into the Nordegg River the two decided to take out their pocket fishermen (folding fishing rods that can be carried in a pocket or glove box) in order to catch some fresh fish for lunch. As the two men sat, they were startled by a loud animal screeching noise coming from a clearing not too far away. Curiosity set in and the two men put down their pocket fishing rods and picked up their rifles and went to investigate. As soon as the men entered the clearing, they both saw a strange animal running toward the tree line about 400 yards away.

The two just stopped in their tracks and didn't say a word to each other. The creature was in sight for about five seconds, and the two men didn't have time to think about raising their rifles, let alone actually doing so. Miles told me the simply turned to Wayne and asked, "Did you see that?" Wayne did not reply but just continued to stare at the now-empty clearing. After about twenty seconds, Miles repeated the question. Wayne just nodded his response.

I learned of this incident from Miles a few years later when he came across my ad in the Calgary papers. We arranged an interview at his farm, which was just east of Calgary. He is a quiet man, not the type to make up wild stories and he also requested that I do not reveal his real name to anybody. This I agreed to do.

Q: "What date did this take place?"

A: "Around Thanksgiving of 1982, October, 1982."

Q: "Was it at night or day?"

A: "It was during the day."

Q: "What time of day was it?"

A: "It was about 10 o'clock in the morning."

Q: "Describe the area in which it took place."

A: "It was definitely elk country, a lot of hard ground and low-lying marshes. When I spotted him, he was by a creek, so it was pretty marshy in there, but there were also hills."

Q: "What distance would you estimate you were from the creature?"

A: "Maybe about 400 yards."

Q: "What was your first reaction?"

A: "I couldn't believe what I was seeing."

Q: "What was it doing?"

A: "It was running."

Q: "Did it run on two legs?"

A: "It was running on two legs."

Q: "Did it ever go down on all fours?"

A: "No, I never saw it do that."

Q: "Was it hairy?"

A: "Oh yes."

Q: "What color was it?"

A: "The color of a moose."

Q: "How tall would you estimate this creature to have been?"

A: "I'm guessing...seven or eight feet maybe."

Q: "What would you estimate its weight to have been?"

A: "400 pounds."

Q: "Did you see any facial features?"

A: "No."

Q: "Could you describe the arms?"

A: "No. It was running—just hairy."

Q: "Could you see if it was male or female?"

A: "No."

Q: "How long did you see this creature?"

A: "About five seconds."

Q: "Did it ever make any noise?"

A: "Yeah, it did. I heard a noise prior to seeing it, and the noise was...you know how lonesome a wolf sounds and you know how lonesome a loon sounds? Well, if you put them together, you come up with a screech, I guess."

Q: "Like a scream?"

A: "Yes. It was low, then it came up high. You know how a moose will make a sound, starting at a low pitch, then going to a high pitch? Quite similar to that, except it was like it was scared."

Q: "Did it see you?"

A: "I believe it did, that's why it was running. But I can't say for sure."

Author's note: Neither man actually saw the creature voicing this screaming noise. They just assume it came from the creature they saw.

Q: "Did you smell any powerful odor before or during this sighting?"

A: "No."

Q: "After the creature moved off, did you check for footprints?"

A: "Yeah, we didn't find any. The ground was too hard."

Q: "Did you report what you saw to the police?"

A: "No."

Q: "Did you report it to anyone other than the police?"

A: "No."

Q: "In your own words describe what happened."

A: "Well, we were hunting, and we stopped by Owl Creek to do some fishing. Then we suddenly heard that noise. We went to investigate and when we went around the bushes into the clearing we saw it running into the bush. Then after thinking about it for a few minutes, we went to see if there were any tracks. There were no tracks. We discussed with each other to see if we both saw the same thing. We never saw it again."

Q: "What was the name of the man who was with you?"

A: "Wayne Farson (not his real name)."

Q: "There were just the two of you?"

A: "Yep."

After the interview, Miles showed me where he saw the sasquatch by pointing out the area on one of his topographical maps he keeps in a large suitcase for hunting. Miles has been farming and hunting all his life and explained there's no way he could have mistaken a bear or some other common animal for the creature he saw that day. When I asked if it could have been a man, or somebody in a costume, he again said no way. He also asked why anybody in some kind of costume would be so stupid as to run by two men with rifles. Additionally, only their wives knew which area they were going to that morning.

We talked for a while and Miles asked me questions on other sightings in the area, and was relieved when I told him that many other people had reported similar creatures in the Nordegg area. As I was getting into my truck to leave, Miles said, "Never in my life have I seen anything like it, and I've never seen anything like it since."

This next report came to my attention when I received a phone call from a Mr. Paul Fergusson who was a reporter for the *Mountain Air Newspaper* in Rocky Mountain House. He told me that a man named Dirk Lundy had called him to report that he and three others saw a sasquatch two days earlier just south of the Sunchild Indian Reserve, No. 202, which lies about one hour's drive northwest of Rocky Mountain House. Mr. Fergusson thought it would be great to do a story on this and since I was known for my research in this area he asked if I would be interested in following up on it. Of course, I said I would.

The next day Paul phoned me back to inform me that as soon as he mentioned to Mr. Lundy that I was told of his encounter and was going to follow up on it, he suddenly decided to clam up. "He doesn't want to talk about it now," Paul told me. I wondered why Mr. Lundy was willing to talk to a newspaper reporter but not me. Maybe this was an attempt at a hoax. I tried phoning Mr. Lundy a few times but he did not return any of my calls. I might have let the matter drop at this point, but Paul did give me the name of one of the other men who saw the creature, a man named Ronald Schulz. I decided to give him a try.

Unlike Mr. Lundy, Mr. Schulz was willing to talk, and invited me out to his home in the morning and told me he was going out hunting that afternoon to the same area and he invited me along so he could show me the exact spot. He also told me the other two witnesses were his sons, Ron Jr., age eighteen, and Robin, age sixteen.

The three of them were hunting with Dirk Lundy on the morning of October 21, 1989. They hadn't had much luck that morning, and the four were taking a break by their truck, enjoying some coffee and hot chocolate from thermoses. The truck was parked along the side of a forestry road, about 400 meters south of the bridge that crosses the Baptiste River. Suddenly, one of the boys drew their attention to something he spotted moving on the north side of the bridge. All four watched as a large, upright creature came out of the tree line on the north side of the bridge. It walked to the road side, seemed to be deciding whether or not to cross the road, turned around and took a few steps back toward the tree line. Then it stopped and again came out to the road side. At this point the creature seemed to spot the four men watching it, for it looked in their direction, then it ran back into the trees. It was not seen again. It was in sight for about fifteen to twenty seconds.

While it was in sight Dirk Lundy brought his rifle up and was watching through his scope, though he did not shoot at it. All four then piled into the truck. They crossed the bridge and slowly drove by the spot where the creature had been, though there was now no sign of it. Later they returned with three other men to have a look around in the bush where the fallen trees had the bark torn off by something that seemed to be feeding on termites. Whatever had torn off the bark didn't leave any claw marks. Ronald was of the opinion that the creature he saw was eating the termites, though none of them actually saw it do this.

One week after this incident took place I was sitting at Mr. Schulz's dining room table for a full interview.

Q: "Would you state your full name please?"

A: "Ronald J. Schulz."

Q: "Where did this incident occur?"

A: "On the Baptiste River, south of the Sunchild Indian Reserve."

Q: "What date did this take place?"

A: "Saturday, the 21st of October."

Q: "Was it at night or day?"

A: "Daytime."

Q: "What time?"

A: "About 10:30 in the morning."

Q: "Describe the area in which this took place."

A: "Lots of bush on both sides of the road, heavily treed, right along the main road that goes through here."

Q: "What distance would you estimate you were from the animal when you saw it?"

A: "I would say between 600 and 700 yards."

Q: "What was your first reaction?"

A: "I thought it was a man coming out of the bush when I first saw it. Then after we thought about it for a while, we kind of got together on our thoughts and...well, no man would do that, eh. I mean the way he came out, and the way he walked and the way he went along the bush and so on and so forth."

Q: "What was it doing?"

A: "Well it looked to me like he wanted to cross the road. He came out of the bush, and he stopped before the road, and then he turned to his left and kind of angled back toward the bush, and then went along the edge of the bush, and then he turned to his right, came back out again about three or four steps. It was then I think he saw us, or heard us or whatever. I don't know, but he turned around and ran back into the bush and was gone."

Q: "Did it stand and walk on two legs?"

A: "Yeah, just two legs, that's all it walked on."

Q: "Did it ever go down on all fours?"

A: "No."

Q: "Was it covered in hair?"

A: "Nearest as I could see, yes. It was black from top to bottom."

Q: "How tall would you estimate this thing to have been?"

A: "I would say at least seven feet tall."

Q: "What would you estimate its weight to have been?"

A: "I would...I'm guessing...250 to 275 pounds—somewhere around there."

Q: "Heavier or lighter than a man?"

A: "Heavier."

Q: "Did you see any facial features?"

A: "No, not from that distance, no."

Q: "Could you describe the animal's arms?"

A: "Well, his arms, well at that distance, they looked fairly big. When he stood, he stooped forward a bit with his upper body, the arms hung down, and when he ran, the arms hung the same way. Flopping in the wind, you might say."

Q: "Could you tell if it was male or female?"

A: "No."

Q: "How long did you see this thing for?"

A: "Oh gee...I would say maybe fifteen or twenty seconds. It wasn't that long. I would have liked to have seen it for a little longer."

Q: "Did it make any noise?"

A: "No."

Q: "Did it see you?"

A: "Well, like I said, I don't know if it saw us or heard us. We were all standing there quiet. Dirk had the scope on him the whole time as soon as he come out of the bush. But I didn't get my field glasses on him quick enough."

Q: "Did you smell anything?"

A: "No."

Q: "After it was gone, did you check for footprints?"

A: "Yes we did."

Q: "Did you find any?"

A: "No."

Q: "Did you report what you saw to the police?"

A: "No."

Q: "You were not the only witness?"

A: "No. There were four of us."

Q: "In your own words, describe what happened."

A: "Well, what happened is that we had gone hunting early in the morning. We had gone into the bush about a mile back. From the road that we were on in and around the bush, and then we all met at the river. The Baptiste River, and after we all got together there, we split up again. Myself and my son Robin walked up a cut line and came back out on the road. Then my other son, Ron Jr., he and Dirk went further north. They had worked their way up to another cut line, and Ron Jr. came out behind us. Dirk followed a game trail, right up a river, and he came out onto the road, and I in the mean time had sent Robin back up the road to get the truck. When the truck came down, we were all standing around having a cup of coffee.

"I'm not sure which one of the boys saw it first, but it came out of the bush. It just walked out of the bush just like a hunter would. I didn't know what the hell to think, because he walked out there and I seen him walk out and he stopped, was standing there and then turned and walked back to the bush. Dirk had his rifle there so he put the scope on it and watched him. It walked along the edge of the bush a little bit, then it turned and came out to the road again. It looked like he wanted to cross the road again. Now, I don't know if he stood there for a few seconds then turned and ran back into the bush. When it had gone, we drove up to the spot real slow. Dirk thought it might be some old trapper. We drove by real slow, went up the road a bit, turned around and came back; we didn't see any tracks or cars further up the road. I looked up the creek, didn't see anything, so we thought we would go back and get the rest of the guys.

"There were three more hunters with us, so we went and got them. We came back down again around 11:30, we were telling the other guys about this, and one of them said, 'Well maybe it was a sasquatch.' We got down there, literally got down on our hands and knees where this thing had walked, and we didn't find nothing. We then went into the bush and checked along the river to see if we could find any tracks or anything along the edge of the river. The only thing that we did find was this log or dead tree. There were two places on it, I guess it would be four or five feet long in places, two places where the bark had been peeled off the log. Bears usually leave claw marks, eh, but I couldn't see any marks."

Q: "You said that Dirk had his rifle sights on it?"

A: "He had his scope right on him."

Q: "He didn't shoot?"

A: "No, he did not shoot."

After the interview, I followed Ronald to the site. He was planning to spend another week hunting a little further west. But he agreed to show me the site. It was October 28. The first snow fall of the season had occurred during the week between the sighting and when I was there. It was fairly warm again and the snow was melting. The area fit his description to the letter. While we were there, he brought out his rifle. I asked if he was just being ready in case we came across a deer. He told me that was a good idea, but the reason he had his rifle was in case we saw the creature again and it turned out to be aggressive. I took his picture standing on the spot where the creature had come out of the bush, holding his rifle over his head

Ronald Schulz, standing on the spot where he and his two sons, as well as Dirk Lundy, watched a sasquatch come out of the trees. He is using his rifle to demonstrate how tall the creature was.

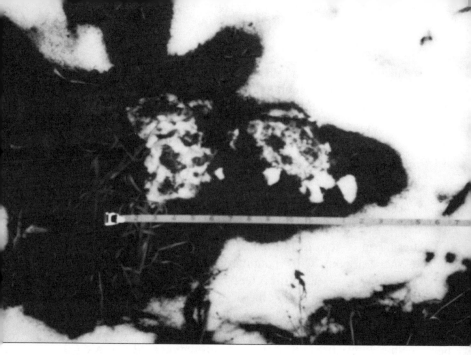

Above: The impression on the river bank which Ronald thought could have been a footprint left by the creature. I'm not so sure.

Below: A fallen tree where it appears something had been feeding on termites. There were no claw marks or beaver teeth marks left on these trees. However, I was able to remove bark from another tree without much effort. The fallen tree was about 100 feet in the tree line, where the four men saw the sasquatch run.

Photos: T. Steenburg.

to indicate how tall he thought it was. I asked if the head of the crea- ture was where the rifle was. He said, "No, the shoulders, I can't reach that high for the top of the head."

About 100 yards into the bush we found the dead trees. One of the trees had bark removed from two places. I examined them, and I found no indication that a bear had been clawing at the tree. Bears are plentiful in this area and are known to claw at trees, leaving behind scratch marks from their claws. No claw marks were detect- ed here, nor were there any beaver teeth marks on any of the trees. However, I did try to pull some bark from another fallen tree near- by, and the bark did come off without too much effort.

While I was at the trees, Ronald called for me to check out the riverbank. He pointed to something in the mud and asked, "Is that a track?" There was an impression in the mud that did seem to resem- ble a footprint somewhat. However, there were certain things wrong with it. There were only four toes. Also, the length of the track was just under twelve inches. A little small for a creature as large as what was reported. I thought what we had here were two deer or elk tracks that were side by side. What made the heel mark, I don't know. Part of me wonders what Ronald was doing while I was studying the dead trees. The other dimensions of this impression were 7 inches across the toes, 5 inches across the ball, 6 inches across at the heel. The toes were imbedded about 2 inches deep in the dirt and the heel was only 1 inch down. The middle of the foot was only an inch deep in the dirt. There was snow in this impression so I don't really think Ronald made this print while I was at the trees. The print was pointing down hill, so this could explain why the toes were deeper than the heel. I do feel that these four men saw something on October 21, 1989. However, certain things about this case still bother me.

Could Ronald and his two boys been the victims of a hoax? I'm still bothered by the fact Dirk Lundy was willing to tell all to a reporter, but changed his mind when he got word that I was going to investigate. Also Ronald told me that he, the two boys, Dirk Lundy, as well as three other men had searched this area with a fine- tooth comb, the day of the sighting and did not find any tracks. Then one week later Ronald and I find one in a spot where I don't think they would have missed. It could well be that this is not a sasquatch track at all, and my first conclusion is right. This was made by either a deer or elk, and the fact it seems to resemble a sasquatch footprint was just coincidence. Also Ronald's two sons were not at home when I went to interview him. I would think Ronald would make sure they were around to tell their side. All these doubts aside, I think this report has a ring of truth to it. However, I could be wrong.

This report is the last I will put in this book from the Nordegg to Rocky Mountain House area of the province. All of these encounters occurred just a little north, or a little south of Highway 11 (David Thompson Highway), as well as right on the highway in some cases. I do have other reports in this area, but they lack detail in my opinion.

It is of note that at least three reports from this area state that the creature was unusually tall. Most reports of this creature in B.C. and the Pacific Northwest of the U.S.A. state the sasquatch stands between seven and eight feet tall. Most reports from Alberta also place the creature at seven to eight feet tall. In the Nordegg-Rocky Mountain House corridor again the most common height reported again is between seven and eight feet tall. However, there have been a certain number of reports from this area that claim the creature was much taller; thirteen feet, fourteen feet and fifteen feet have been reported. I always assumed that the witness, in the excitement of the moment, just overestimated the creature's height. I still think this is probably the case. However, I have found it interesting that all of the extreme height reports have come from the same general area—why? All reports of extremely tall creatures have been within about 100 miles of the Big Horn Dam; there are many normal height reports (seven to eight feet) around there as well. The last extreme height report I have on file happened in 1984. There have been none since.

Could it be that Alberta had a few really tall creatures that lived out their lives in this area and have since died? We know that humans occasionally grow very tall. The NBA is full of seven-footers. It may very well be that every now and then sasquatch will grow much taller than the rest. However, I still have very hard time accepting a height of fifteen feet—nine feet maybe, but fifteen feet?

**A 6 foot two inch man,
in comparison with an 8 foot high Sasquatch.**

I still think that the witnesses overestimated the height of the creature they were looking at. But it is interesting that all of these over-estimations have occurred in the same general area. It is even more interesting that the estimates have now stopped.

I remember during a conference in Pullman, Washington, a young woman asked me, "How come there are no reports from the National Parks?" Well, there have been. Along the Alberta-British Columbia boundary, especially the southern half of the province, there are very few places where you can drive from one province to the other without passing through a park. In the south of the province lies Waterton Lakes National Park. As you head north along the boundary, there's Peter Lougheed Provincial Park. Next to that is Kananaskis Provincial Park, then Kootenay National Park. Just east of Kootenay lies Canada's oldest park, Banff National Park, then Jasper National Park. To the west Yoho National Park, again to the north Mount Robson Provincial Park. All these national and provincial treasures comprise a wilderness area almost as big as the entire land mass of England. One would think that this area known for its grizzly bear population would harbor sasquatch as well. This does seem to be the case.

One of the earliest accounts on record of strange footprints was made by the famous explorer David Thompson, who wrote in his journal about strange footprints his party came across near the present site of the town of Jasper in 1811. The town of Jasper is right in the middle of today's Jasper National Park. The entry is dated January 7, 1811.

> Continuing our journey in the afternoon we came on the track of a large animal in the snow, about six inches deep on the ice. I measured it: four large toes each of four inches in length. To each a short claw; the ball of the foot stuck three inches lower than the toes. The hinder part of the foot did not mark well, the length of fourteen inches by eight inches in breadth, walking north to south and having passed about six hours. We were in no humor to follow him. The men and Indians would have it to be a young mammoth and I held it to be the track of a large grizzled bear, yet the shortness of the nails, the ball of the foot, and its great size was not that of a bear, otherwise that of a very old bear, his claws worn away. This the Indians would not allow.
>
> (Thompson, David. *Narrative of his Explorations in Western America*, 1784–1812. p.36)

Maybe they were tracks of a bear. Although I feel Thompson and his party saw many bear tracks as they crossed the Rocky Mountains. I cannot help but think that for a set of animal tracks to gain special attention from a man like David Thompson, they must have appeared very strange indeed. All travel in 1811 was done on foot or horseback. For him, noting animal tracks as an unusual occurrence, he must have been either exceedingly bored or going a little mountain happy, or he found a set of tracks with such rare qualities that they demanded special consideration. The following autumn, while he was passing through the same area on his return trip he mentions the tracks again.

> I now recur to what I have already noticed in the early part of last winter, when proceeding up the Athabaska River to cross the mountains in company with the men and four hunters, on one of the channels of the river we came to the track of a large animal which measured breadth by tape line. As the snow was about six inches in depth, the track was well defined and we could see it for a full hundred yards from us. This animal was proceeding from north to south. We did not attempt to follow it, we had not the time for it and the hunters, eager as they are to follow and shoot every animal, made no attempt to follow this beast, for what could the balls of our fowling guns do against such an animal. Report

from old times had made the head branches of this river, and the mountains in the vicinity the abode of one or more, very large animals, to which I never appeared to give credence; for these reports appeared to arise from the fondness for the marvelous so common to mankind, but the sight of the track of that large beast staggered me, and I often thought of it, yet never could bring myself to believe such an animal existed, but thought it might be the track of some monster bear.

Mr. Thompson here reminds me of some anthropologists who were at first very impressed with footprints they looked at during the late 1950s and early 1960s, but later talked themselves out of it because they had always assumed that such a creature couldn't possibly exist. Some things never change I suppose.

The late Roger Patterson received a letter from a Mr. Gerald Martin, who told him about a strange creature he and his family watched on a ridge, while on vacation in Jasper National Park in August, 1968. The family was sightseeing on the Columbia Ice Fields. Gerald Martin was the first to notice the creature, which seemed apelike in appearance. But all agreed it was too tall and moved too fast to be a man. However, the creature was very far away, so no detail was observed.

I have a letter, which was forwarded to me by John Green, from a Drayton Valley man who requested that his name not be revealed, concerning a strange animal he and two other men encountered in the Lake Ribbon Creek area of Kananaskis Provincial Park, in September, 1969.

John Hawker (not his real name) was with two other men, whose last names were withheld from the letter.

Dear Sir,

Let me begin by saying that what I'm going to tell you is true. My first brother-in-law once had the idea of doing spare time prospecting in the Kananaskis, Lake Ribbon Creek area of Alberta. Approx four years later, while I was working for an oil co. In Calgary, some of the men in my dept, were talking of old mining operations. I remembered Ribbon Creek, told them of the place, and we set a date to go and take a look around. I'm sorry but the exact dates escape me.

Al _____, & Gordon _____, and myself, drove from Calgary on a Saturday morning. I believe it was around the middle of September 1969. We agreed never to tell anyone what happened for reasons to follow. We arrived at Ribbon Creek, approx 8 or 9, on the Sat morning. Spent the morning just looking for old signs of mining. At noon we built a fire to heat up some canned food. We had been sitting for about hr, when Al said, "Hey what's that?"About 100 yards away we saw what looked like a gorilla sitting on its haunches, watching us. Well I wanted to head for the car and so did Al. But Gordon said, "Wait a minute,"in sort of a whisper. The animal sat there for about five minutes, then it got up and sort of chattered its teeth, like it was cold, and also moved its arms in an up and down movement. After the animal had gone we walked to the place where it was sitting. We could not make out any tracks, but we all agreed that it stood at least 7' to 8' tall. On Sunday I talked my landlord to drive back with me to ribbon Creek. We didn't make it until dusk. On the road to the site we noticed some fresh tracks on a shale slope beside the road. My landlord found a good track, presumably of the animal. He took three photographs and we tried to measure it, though we had no tape. I do hope this incident might help your studies. I wish to remain anonymous. I have never mentioned this incident to any-one and I hope that you will respect my wish. Also if you wish, I can speak to a man living out in bush country, about 35 miles west of Drayton Valley. This man has lived in the bush since childhood and he claims to know quite a lot about these animals, and has observed them many times. I am a born cynic and if I had not seen one with my own eyes I would never have believed they exist. Enclosed is a sketch of the animal we saw, and a diagram of the footprint. Please excuse the spelling.

Yours truly

John Hawker

The sketch he sent seems to indicate that the creature they encountered was a female, due to the fact it has large, droopy breasts. Also a footnote on the sketch indicates that the creature's head was sort of pointed. The creature stood between seven and eight feet tall and was covered with long hair that was brownish in color. Whether this means it was light brown or a brownish gray color is unknown. The creature's arms are described as longish. The sketch of one of the footprints shows five toes, and a length of fourteen, fifteen or sixteen inches.

Remember, Mr. Hawker said that he and his landlord did not have a measuring tape with them, so they were guessing. The heel seems strangely wider than the forefoot. This could be a simple matter of Mr. Hawker not being a very good artist. None of the photographs taken of the tracks by his landlord have reached my files, and I have never seen them to compare with the sketch. As for the man who has spent most of his life in the woods, was said to know a great deal about these creatures and had observed them many times—this man has not been identified.

The town of Banff, located in Banff National Park, is probably Canada's most famous wilderness resort town. Millions of tourists, from hikers to skiers, arrive each year. Recently the growing popularity of the town has begun to create concern over development. Banff still has elk in the streets, which causes some problems during the rut (annual period when elk are in heat), when tourists come too close. In 1980, panic almost gripped the people of Banff, when a huge grizzly bear mauled three victims over a period of a week on some hiking trails just outside the town limits. One of the men died of his injuries. When park wardens finally shot the bear they were amazed at its huge size. These attacks marked the end of open pit dumps in the national parks. The hide of this bear is now on display in the town's wildlife museum, located on the Bow River. As would be expected, there have been a number of reported sightings of sasquatch around the town as well.

I was contacted by a Calgary man, Ray Shoeman, who told me that as a boy he was in the family car with his two brothers and father driving west in Banff National Park. They had just passed the turn off into the town and were passing Vermilion Lakes, when Ray saw some strange animal near the opposite shore. There was still ice on the lake and snow on the ground. He pointed out the creature to his brothers who became very excited and called to their father to go back. He wouldn't, scolding the boys for the commotion they were causing. Mr. Shoeman came to my home for an interview.

Q: "State your full name please."

A: "My full name is Ray Herbert Shoeman."

167

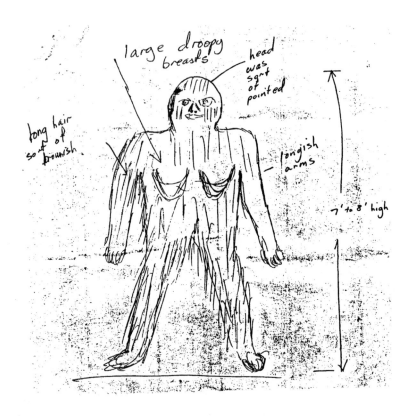

large droopy breasts

head was sqrt of pointed

long hair of sort of brownish.

longish arms

7' to 8' high

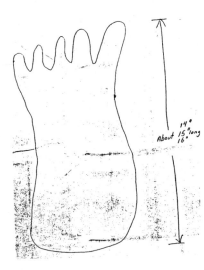

About 15" long 14° 16°

Above: Sketch drawn by John Hawker (not his real name) of creature he and another man encountered in the Kananaskis-Ribbon Creek area of Alberta, in September, 1969. It would appear the creature was a female, due to the fact it had large, droopy breasts.

Left: Sketch of footprint drawn by John Hawker (not his real name). These tracks were found by Mr. Hawker, who returned to the site with his landlord a few days later. These tracks were photographed by John's landlord at the time. However copies of these pictures have not reached my files, nor have I ever seen them. If the drawing is correct, the heel seems unusually wide in comparison with the forefoot.

Q: "Where did your sighting take place?"

A: "Vermilion Lakes, just outside of Banff, Alberta."

Q: "What date did this take place?"

A: "Spring time, 1977."

Q: "Was it at night or day?"

A: "It was during the day, it was sunny out, very clear, probably just few clouds out, and I could see that there was no markings on the snow whatsoever, the snow was completely unmarked."

Q: "What time of day was it?"

A: "About midmorning."

Q: "Describe the area in which this took place."

A: "It was clear, completely clear, it was right out in the open."

Q: "What distance would you estimate you were from the animal when you saw it?"

A: "I would guess probably 200 meters, maybe three city blocks."

Q: "What was your first reaction?"

A: "Well I glanced...well I normally watch over the mountains to see, you know, if anything is different about the mountains today than a few weeks ago, or years ago. I spotted something on the lake, I turned away then it struck me that it was possibly a bigfoot that I had seen. So I quickly turned back in a matter of seconds. Then I got really excited. I pressed my face against the window, and two of my brothers were in the car with me. My father was driving the car, my mother was in the front passenger seat, so I grabbed my brother and hauled him over to the window and we all looked out the window, and we all saw the same thing. I saw more of it than they did because I was sitting on the side, the driver side of the vehicle, and we were traveling west on Highway 1, so the lake was on the driver's side. So I got a better look at it than my brothers did, but they did spot something big and black on the lake."

Q: "What was it doing?"

A: "It was walking upright, like a man. It couldn't have been...I've ruled out a cross-country skier because, as I said, there was no tracks on the lake period. I looked for tracks, it's kind of vague, what I saw behind it, but it looked more like footprints behind it, than ski tracks. The movement of this...bigfoot...was no where near what a cross-country skier would do, okay? For a cross-

The Blackstone River, northwest of Nordegg, Alberta. Located in the middle of Alberta's main area of reported sightings.

Photo: T. Steenburg, 1994.

country skier to break through fresh snow or untracked snow, he would be taking very small strides. But this creature was taking very long strides, walking upright like a man."

Q: "It was walking upright?"

A: "Yes."

Q: "Did you ever see it go down on all fours?"

A: "No, and I saw it take about ten steps."

Q: "Was it covered with hair?"

A: "It appeared to be hair...it was almost a shiny coat, the sun was kind of reflecting off it a bit. I know something hairy wouldn't shine in the sun, but whatever it was, it appeared to have a very clean coat. I've seen bears with shiny coats, and bears are one of the worst keepers of their own fur for animals around, you know, dogs are the same, cats are very clean. I guess you can say that this thing had a coat like a cat, not really shiny, but the sun kind of reflected off the coat."

Q: "What color was it?"

A: "And it had a bit of a blue tinge to it, because of the sun shining off of it."

Q: "How tall would you estimate this creature to have been?"

A: "Oh my...it would have to have been, if I compare it to a bear or some other animal...or even a man, I would have to say somewhere around 400 pounds, 350 to 400 pounds at least."

Q: "Did you see any facial features?"

A: "No, not really. It was kind of looking around. It was turning its head. But we were too far away to see any facial features."

Q: "Could you describe its arms?"

A: "Very, very long, and they were swinging, it was probably a two-meter swing from front to back."

Q: "Could you tell if it was male or female?"

A: "No, but my guess would male."

Q: "How long did you see this thing for?"

A: "Between ten and fifteen seconds in total."

Q: "Did it ever make any noise?"

A: "I couldn't hear anything because we were in a moving car."

The sasquatch display at the Natural History Museum, located at the Clock Tower Village, 112 Banff Ave., Banff, Alberta. This display consists of a life-size statue of a sasquatch, and is well worth seeing. I agree with the general body proportions, however the skin color of the face and hands is wrong.
Photo: T. Steenburg, 1988

Q: "Did it see you?"

A: "No. It probably saw all the traffic on the road, but not me in particular, no."

Q: "Did you smell anything before or during this sighting?"

A: "No, again because we were in a moving car."

Q: "Did you check to see if there was any footprints?"

A: "Ah, I begged and pleaded with my father to turn the car around and go back but he refused, telling me that it was just my imagination. I really begged and pleaded, I wanted to go back, but no, I didn't go back to investigate anything."

Q: "Did you report what you saw to park wardens or police?"

A: "No, I didn't."

Q: "Did you report it to anyone?"

A: "No."

Q: "You weren't the only witness?"

A: "I have two brothers who saw it too, but they didn't see it for as long as I did. I probably saw it for at least five seconds more than they did."

Q: "What were their names?"

A: "Robert and Eric."

Q: "What were their ages at the time of this incident?"

A: "That was in 1977 so I was twelve, my brother Rob would have been sixteen. My brother Eric would have been ten."

Q: "In your own words describe what happened."

A: "We were driving down Highway 1, traveling west, past Banff, and there is a hill on the highway which begins just past the Mount Norquay overpass, right close to Banff. There is a hill on the west side of the overpass, and just as we were cresting the hill...let me back up a bit. I was looking around outside through the window of the car, and I was looking south at Vermilion Lakes, and just as we were cresting the hill, I saw this unusual creature on the lake. I thought nothing of it for just a moment, and I looked again and got a very, very good look at what it was...to me it looked like bigfoot. It couldn't have been, well...it had to have been 150 meters (500 feet) away. It was walking upright, right in the middle of the lake, and I had a clear, open view. I couldn't have had a better look from that distance.

"I'm quite sure if there were other cars on the highway that maybe somebody else spotted it also. I can't be sure of that, it was such a long time ago. But if they did see the same thing that I did, then they haven't come forward. I watched this...bigfoot...for as long as I could. We were traveling about 70 to 80 kilometers an hour (50 mph), and the distance where I spotted it and where I lost sight of it was at least half a kilometer (1/4 mi.). The whole thing lasted about ten to fifteen seconds. Where I lost sight of it is where this cliff had been blasted out from years before to put the highway through, blocked my view, and I couldn't see it anymore, until we had come around the corner. I jumped into the back of the car, it was a Cherokee Jeep by the way, I jumped into the back and I watched for as long as I could. My brothers saw the same thing. We could see this black thing out on the lake, in the middle of the lake, walking across it. I know they would both back up my story, because they both saw the same thing."

Q: "You asked your father to stop the vehicle?"

A: "Yeah, well I begged and pleaded for him, come on, let's turn the car around and go back and have a better look! I really begged and pleaded with him, but he didn't stop and he told me it was just my imagination, and not to make such a big issue of it. I really, really wish that we had turned around and gone back. Then I could of really known for myself whether or not it was really bigfoot that was out there. But it definitely not a bear walking along, because a bear can only take very short steps on its hind legs, and they don't walk upright for very long. I don't think it could have been a cross-country skier, because I've never seen a cross-country skier move like that. It could not have been a man because it was too tall to be a man."

Ray Shoeman has been bothered by what he saw, or thinks he saw, since that day in 1977. The highway was still a single lane then, it wasn't doubled until the mid-1980s. I still remember myself the traffic jams as the work was going on. So it was too bad that their father didn't turn around and go back for a second look, if he had his son might have been spared years of wondering. This part of the TransCanada Highway is very busy indeed, so I think that other

173

motorists must have seen the same thing that day. But so far, nobody has come forward.

There was a similar incident twelve years later, only a few kilometers east of the town of Banff in 1989. Again the creature was seen by a family driving west in their car. Terri Pacileo, who owns and runs a very tasteful art store in Calgary phoned me in 1994 to tell me about the strange animal she and her two children saw. She agreed to a full interview, so I made arrangements to meet her at her store a couple of days later. When Barbara and I arrived at the store, Terri welcomed us with tea, and showed us to a back room where the interview was to be held.

Q: "Would you state your full name please."

A: "My name is Terri Pacileo, Terrica Laydel Pacileo."

Q: "Where did this incident occur?"

A: "Just outside of Banff, inside the park."

Q: "What time of year was it?"

A: "It was late summer."

Q: "Do you remember the month?"

A: "It was late July."

Q: "Do you remember the date?"

A: "No, I didn't mark it down in my book. It was at least five years ago."

Q: "Was it night or day?"

A: "It was daytime. Late afternoon."

Q: "What was the weather like that day?"

A: "Good. Very nice weather, clear."

Q: "Describe the area in which this incident occurred."

A: "It was a very wooded area. There was a grass area off the highway, thick trees. A wooded area, and that's where we saw it in the wooded area."

Q: "Were you in a car?"

A: "Yeah, we were driving."

Q: "Were you driving toward Banff, or east toward Calgary?"

A: "Toward Banff. We were going to B.C."

Q: "Was it on the left-hand side of the road or the right?"

A: "Left-hand side."

Q: "What distance would you estimate you were from this thing?"

A: "I don't know...maybe thirty feet."

Q: "So it was right next to the road?"

A: "No. There was a little stretch of grass. Seeing the size of him, I can't imagine that we would be more than thirty feet, cause we saw him, very tall, very clear. About thirty feet I think."

Author's note: I have to conclude that the creature was in the tree-covered highway divide which separates eastbound lanes from westbound. This divide is almost 100 yards across in some places, filled with mature trees, and westbound traffic cannot see the eastbound lanes. In other places it's only about thirty feet wide. If this sighting occurred just before the turn off into the town of Banff itself, it is possible they saw the creature in the trees across the eastbound lanes, since in this area only a concrete divider separates east and westbound traffic. If the incident occurred here, then I feel Terri has underestimated the distance from the car to the animal they observed, due to the fact the eastbound lines alone have three line lines of traffic and the highway alone is more than thirty feet wide.

Q: "What was your first reaction?"

A: "My first reaction was amazement. I was really excited because both my children saw him. So we could compare notes. We all saw the same thing."

Q: "What was it doing?"

A: "It was running. Just running."

Q: "Away from you or toward you?"

A: "Parallel with us really."

Q: "Along the road?"

A: "Yeah. He wasn't like right next to the road. He was closer to this wooded area, but as we were driving we were moving next to him."

Q: "So it was in the trees?"

A: "Yeah. Well not really. He was very visible just outside that wooded area."

Q: "Did it stand and walk on two legs?"

A: "Two legs, yes."

Q: "Did it ever go down on all fours?"

A: "I didn't see it go down, no."

Q: "Was it covered in hair?"

A: "Lots of hair."

Q: "What color was it?"

A: "It was an orangy brown color."

Q: "How tall would you estimate it was?"

A: "It was taller than a man. It would have to be taller than a man to see him from that distance. He had a little bit of a stoop, so I wonder if it was...eight feet. It could have even been higher because, you know if it was erect?"

Q: "It was stooped?"

A: "Yeah, just a little bit of a stoop in its posture when it was walking."

Q: "What would you estimate its weight to have been?"

A: "Heavier than any man I've ever known. Very heavy...I'm not good at guessing weight at all. Two hundred pounds would be nothing, 300 pounds would be nothing."

Q: "Very heavy?"

A: "Very heavy, yeah, oh yeah."

Q: "Did you see any facial features?"

A: "No, but the thing did remind me of a man."

Q: "When you saw it, was it from the back?"

A: "The side."

Q: "Could you describe the arms for me?"

A: "I just remember lots of hair from here (indicates long hair hanging under arms), this is where the longest part of the hair was. His hands were down. Like he reminded me, kind like a large ape, but the way he was running reminded me of a very tall man."

Q: "A human stride?"

A: "Oh he had a lengthy stride. A very lengthy stride."

Q: "Could you se if it was male or female?"

A: "No."

Q: "How long did you see this thing for?"

A: "Oh it would have been for a very fleeting moment, because we were driving about 100 kilometers an hour. He was going fast because we saw it, like my children saw it first, then as I turned I saw it clear enough that it just stamped in my mind. But how long we saw it for? How long I saw it for? It was awhile because we were going parallel with it."

Q: "Did you pass it?"

A: "Yeah, we passed it. It didn't go into the bush. My husband is Italian, we did try to get him to stop, but he wouldn't like he just kept going."

Q: "Your husband was in the car too?"

A: "Yeah."

Q: "Did he see it?"

A: "No. He thought we were crazy. It's just a bear! I said let's go back, but of course he wanted to get to our destination, so he wouldn't go back."

Q: "Did it ever make any noise?"

A: "No."

Q: "Did it see you?"

A: "No. I don't think so. There had to have been other cars on the road. But I don't really remember seeing other cars at this point."

Q: "This was the TransCanada Highway was it not?"

A: "Yeah, right on the TransCanada."

Q: "Did you smell anything?"

A: "No, our windows were up."

Q: "I don't suppose you stopped and checked for footprints or anything?"

A: "No. We wanted to, but my husband wouldn't go back."

Q: "Did you report this to anybody?"

A: "No, nobody. We have told friends. My kids have talked about it since. Even though they were young it was etched into their minds. My seventeen-year-old would have been fourteen. My boy is fourteen now so he must have been eleven. Both of them saw it very clearly."

Q: "Were you inside the park boundaries when you saw this thing?"

A: "Yeah."

Q: "In your own words, describe what happened."

A: "Well we were taking a trip, my family and I. My two kids suddenly said, 'Mom, look quick!' I glanced around and I saw what I first thought was a guy running, and then I thought it was a bear, then I knew it wasn't either, and that it was upright. It was very tall. We knew later that it was the thing described or called bigfoot or whatever, because that was the only thing that would

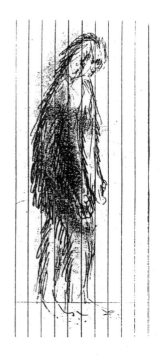

Terri Pacileo drew this sketch (on lined paper) of the creature she and her family saw as they were approaching the town of Banff in July, 1989. She and two of her children saw the creature from the family car as they drove by. She and the children begged her husband to turn around for a better look. He refused.

come to mind. Both of the kids who saw it and I could compare notes and all of us agreed that we saw the same thing. We were just excited by what we saw, it wasn't our imagination, this guy was very hairy, funny-looking creature. He was a wild...apeman of some sort. That's what we saw."

Q: "Your two children saw it?"

A: "Yeah."

Q: "But your husband didn't?"

A: "But my husband didn't We had four kids in the car, but only two saw it. The other two were asleep and one of them would have been too young to remember."

Besides Terri's obvious underestimation of the distance between themselves and the creature. I can find little wrong with her story. At one point in the interview she describes the creature's movement as walking. When she was describing its stooped posture she says at this point that the animal was walking, while at every other point in the interview she says it was running, and running very fast. I suppose she may have seen it do both or maybe she meant running. I did phone her seventeen-year-old daughter and asked her questions about the creature they saw that day. Her daughter confirmed the sighting almost to the letter. The only difference was that she thought it was slightly darker in color than Terri's orangy brown. I showed Terri a still frame from the Patterson/Gimlin film (frame 352 to be exact), and asked her if this was similar to the creature they saw. Their sasquatch was lighter in color, and had longer hair on the arms. Also she though the creature they saw was not quite as broad in the hips, as well as being flat chested. I think from this information we can conclude if it really was a sasquatch they saw in July, 1989, it was probably a male, even though when asked, she had no idea as to the creature's gender.

Again as in the previous report this creature was spotted along the TransCanada Highway near the town of Banff. Since many tourists are driving hoping to see wildlife to photograph, I would think that somebody else must have seen the animal at the same time the Pacileo family did. If so they have not come forward.

The next sasquatch report in my files, which occurred in a national park, is "The Crandell Campground Incident," when four witnesses reported to park wardens that a creature fitting the description of a sasquatch came to their campsite in the early hours of Sunday, May 24, 1988. Crandell Campground is located in Waterton Lakes National Park, in the southwestern corner of the province. This incident is the most interesting and detailed account of a sasquatch that I have yet investigated. In years to come I feel (assuming that a sasquatch has not been brought in) that this case will be listed with other classic stories,such as The Albert Ostman Story, The Ape Canyon Incident, The Ruby Creek Incident and The William Roe Story. All these classic tales have been written about extensively in other books, so I have not repeated them here. The Crandell Campground Incident should be in a chapter on its own. That's what I have decided to do, due to the fact it is so long, all four witnesses were interviewed by myself and the fact I've written about the case in my other two books. So readers who wish to skip it can. However I've never written all the details word for word before so I think you the reader will find chapter ten very interesting indeed.

Another report from Waterton Lakes National Park was only weeks after the Crandell Campground Incident. This incident too was reported to the park warden's office which, because of the Crandell incident, went to investigate at once. Mike Sheen of Lethbridge, Alberta, was on a fishing, camping trip with his boys Justin and Joshua, as well as a third boy named Joey. The four were in a sleeping lean-to of wood construction the park services provide to visitors in the back country, in this case Upper Twin Lake. They had settled in the lean-to early on the morning of July 10, 1988. They then went down a path through the trees to fish in the lake.

After a few hours all four heard a high-pitch screaming noise coming from the area of the lean-to. They couldn't see their camp from the water's edge due to the trees that were between them and the lean-to. they could hear some of their camping gear being thrown about. It was apparent that some animal was destroying their camp. The four boys just stood there in amazement and Mike gathered them together in case whatever it was came in their direction. Whatever was rummaging through their gear was making the screaming noise the whole time. "I've never heard any animal make a noise like that!" he told me.

Mike is an avid outdoorsman and hunter, who spends most of his spare time in the woods. "No bear makes high-pitched screaming noises like that!" he said. After a few minutes whatever was making the noise started moving away. It continued to scream the whole time, they knew it was moving away because the noise was getting fainter and fainter. After about thirty minutes the four slowly went back to the lean-to. They found their backpacks thrown about, their stove and lantern tossed away and some of the sleeping bags had been ripped.

While I was at Mike's home in Lethbridge, Joey showed me his track pants that were in his backpack at the time of the incident. They had been taken from his pack, torn and ripped and discarded about forty yards from the lean-to. None of the four actually saw what it was that tore up their campsite, and they didn't look around for footprints or anything else. Mike had the boys pack up their damaged gear and they left the area hoping to reach Mike's truck before the sun went down.

Later when the warden's office was informed of this incident, they were very interested, especially when Mike told him he didn't think the animal was a bear. They were still abuzz over the Crandell Campground incident a few weeks earlier. The idea a sasquatch was responsible had not occurred to Mike until one of the wardens mentioned it. Mike was taken aback at this question. The warden then told him that such a creature had been seen recently at Crandell Campground, and that's why he asked.

I played some recordings of alleged sasquatch screams on my recorder for them at Mike's home. (They were recorded at the Lummi Tribe Reserve in Washington State in 1975 by a man who was then a police officer, and is today retired. Cha-Das-Ska -Dim, Which-Ta-Lum is his Lummi name; he was known back then as Kenny Cooper. He recorded on his police mike the screaming, or shrieking noise of a sasquatch that had been frightening residents over a number weeks.) When I played the tape for Mike and the boys, all agreed that the noise was the same as the one coming from whatever it was that destroyed their camp. However, the fact remains that none of the four saw the animal in question, therefore we can't say for certain that a sasquatch was to blame.

Like the Mike Sheen report from Waterton Lakes National Park, there have been other incidents in Alberta of campers being frightened by some unseen animal lurking just out of visual range. The scenario almost always seems the same. People relaxing after the day's activities, when some creature, at which the witnesses never seem to get a good look, appears in the trees just beyond the glare of their lamps or campfire. The nervous campers are left shaking and wondering what kind of creature this was. Almost never do the witnesses actually see the creature that was making all the noise. So nobody can really say that it was sasquatch that was responsible. However, it has been reported from eye witnesses that the sasquatch does make a variety of vocalizations. Everything from grunts to barks, as well as whistling noises, have been reported. Sasquatch have also been said to emit very loud and high-pitched screaming or screeching noises. Some people report the animal making such a scream, and then hearing another creature answering far off in the distance. Perhaps the sasquatch uses this method to contact its own kind over great distances—like wolves or coyotes howling. Again I must point out that in such cases, unless the person actually saw the sasquatch and then heard the vocalization coming from the creature as he watched, these observations are of no real use.

I received a phone call in October, 1988, from a Jim Kobylansky, who told me of an incident that happened while he was out camping with three companions in September, 1988. They were in the back country. Jim prefers this to public campgrounds. They had hiked into this area all day, from Strawberry Campground in Kananaskis Country. The four had picked a spot in a small mountain clearing to set camp. At approximately 9:30 P.M. the four were sitting by their campfire, talking and joking around, when they suddenly heard a strange animal noise coming from the trees. Whatever it was seemed to be circling the campsite, staying just far enough inside the trees in order to avoid being seen. Jim had worked as a backcountry guide for many years, and he told me he never heard anything like this animal before. He described the noise as sounding like "ooo-eee-ooo-eee-ooo-eee." It wasn't really a scream or howl. Whatever it was continued to circle the camp for about thirty minutes. One of Jim's companions threw a lot of small wood on the fire to increase the light. The fire flaring up may have had an effect on the intruder, for the sounds for an instant seemed to indicate distress.

"I don't know how to describe it, but when the fire got brighter, the tone from the creature changed for a few moments." Soon the sounds went back to their original pitch. "It was obvious that something didn't like the fact we were around," Jim said. It was now apparent that whatever it was it was leaving. They continued to hear it, but further away each time until they no longer heard it at all. Nobody got any sleep that night, and the four packed up and left when the sun came up. I played alleged sasquatch screams for Jim to see if they were the same. None were, as far as he was concerned.

Again, it must be pointed out that none of the four saw what was making the noise outside the campsite that September night in 1988. So there's no way to know that it was a sasquatch.

The second report of a strange, unidentified animal making noises is especially annoying to me. It occurred at Winchel Lake, about two miles south of Water Valley, Alberta, on June 4, 1992. The reason I find it a bit annoying is because I was living in Water Valley at the time. As a matter of fact, on the night in question I was smoking my pipe on my porch working on the manuscript for my second book, *Sasquatch/Bigfoot: The Continuing Mystery*. A Mr. Gary Rieberer was camping along the north shore of this small lake when he heard the silence broken by a loud animal scream that seemed to come from the trees on the south shore. The noise lasted about a minute each time, and he heard it about half a dozen times,

Max Rieberer took me to his brother's campsite on Winchel Lake, a few miles west of Water Valley, Alberta. Here on June 4, 1992, Gary Rieberer reported strange screaming noises coming from the opposite shore.

the first was at about 9:30 P.M. He described the noise as something between a howl and a woman screaming.

"It must have had a very big chest to make such a cry," he told me over the phone two nights later. "I was on edge the rest of the night." The next morning he told his brother, Max, about what he had heard. Max suggested to him that maybe a sasquatch was making the noise. Gary never thought of that, and told his brother that he never saw what was making the noise. Max then asked, "Do you mind if I tell that sasquatch guy in Water Valley?" He had apparently heard me on Charles Adler's radio show, *Hot Talk*, discussing the sasquatch mystery. Gary said sure, but he wasn't going to come along with him because he never actually saw anything.

The first thing I remember about the case was being outside cutting firewood when a jeep suddenly turned in my driveway and this pleasant, soft-spoken gentleman stepped out and asked me if I was the sasquatch hunter. I chuckled to myself and told him yes, I research reported sightings of this animal. He then went on to tell me about the sounds his brother heard not two miles away two nights before. He wanted to show me the area and I saw no harm in taking look, though I knew by now that his brother hadn't actually seen anything. We searched the south shoreline without finding anything. Later that evening I discussed the matter with Gary over the phone. Again I played tapes of screams for comparison. He told me that the screams recorded by Linda Willowford, near Snohomish, Washington, in 1978 and 1979 were very similar to the cries he heard. Coyotes are very numerous around Water Valley, and I have listened to them many a night outside my house. On the night this happened I remember hearing them myself. Gary told me that he too was familiar with the sounds of coyotes, and this sound was quite different. I, myself, never heard anything two miles away.

Strange animal cries in the night make for interesting reading however there is no way to connect them with sasquatch one way or the other.

The Clear Water River area of Alberta, near the town of Caroline has had numerous reports over the years. This area is second to the Nordegg area, which I have already talked about. I received a letter from a Mr. John Kilborn, who read my ad in the Calgary papers. He wrote about a strange animal he saw near the Ram River Gas Plant, west of Caroline in October, 1973.

In the fall of 1973, I was working at the Ram River Gas Plant. I went out hunting one night after work, looking for deer or moose. I was alone. I walked west on a cutline, about 4 miles from the plant. The hills on the cutline were steep, the tops being about 1/4 mile apart. While on a hill top I looked west to the next hill top, 1/4 mile, and there was something standing there very still. It looked the color of a moose or horse. I watched it for about 10 minutes. While I caught my breath. I was very curious as to what kind of animal this was. I was carrying a 270 with weaver 4 power scope. I couldn't identify it using the scope. I considered firing a shot beside it, but changed my mind. I wanted to see it move so I could identify it. I thought of shooting it, but changed my mind. I decided I had enough daylight left to walk to where it was, so I set out. Arriving at the spot where it had been I looked for tracks, checked the bush for hair and looked for droppings, nothing. While walking in between the hills I lost sight of the animal, the hill was too steep. The ground was soft enough that a horse or moose should have left tracks, none were found. I dismissed the idea that it was a Sasquatch because of the roundness of the top of the body. Until I saw a sketch of one, and it looked like what I had seen. I saw your ad in the paper, but put off calling you, perhaps I should have. It stood there watching me for about 10 minutes of so, while I tried to figure out what it was, and what to do about it. I'm glad I didn't shoot it. I don't believe in shooting an animal without good reason, or unless it's a sure shot. I didn't know what it was for sure but I just never forgot about it.

There was more to John's letter, but all he wrote about was questioning the morality of shooting a sasquatch. Just to clarify one thing though, he didn't see my ad in the press in 1973. I was only twelve years old then. He saw my ad in 1981.

In 1985 I was contacted by a Calgary man who told me of encountering a sasquatch on the bank of the Clear Water River, near the town of Caroline, while he and a friend were out hunting in September, 1976. After he told me the story I felt it had sounded familiar, but I couldn't place it. In the mean time I had made arrangements to visit Gary Schmidt at his home in Calgary to interview him about what he thought he saw. Gary was a soft-spoken man, who was now, due to a bad truck accident, forced to wear a back brace. Gary seemed to me a man who was wishing he could live life over and do a few things different. He had spent most of his spare time doing what he loved to do which was hunting and fishing. During a hunting trip with a friend, they both saw a sasquatch. Gary made no bones about what he saw. He then told me that he had sent a letter to Cheam Publishing Ltd. Hoping they would forward it to John Green. Now I knew why Gary's story sounded familiar. I confirmed this after I returned home and turned to page 241 of John Green's book, *Sasquatch, The Apes Among Us*, and there was a small paragraph, telling about a man seeing a sasquatch in the Clear Water River on September 1, 1976. John did not name the river, but described it as fast flowing. Later when I had put my interview with Gary on paper, I sent a copy to John, so he could compare what Gary had told me to what he had written eight years earlier. First let's start with my interview with Gary Schmidt.

"I was hunting rough grouse, with a friend of mine, along the Clear Water River near Caroline. Were walking along when he yelled up to me, he was on the other side walking through this clearing, and he yelled up to me, he says, 'You got apes up in the mountains?' I went running down and I sees this big animal in the river throwing water on top of itself."

Q: "It was in the river?"

A: "Yes, splashing itself. I ran down toward it, and my friend said, 'Are you crazy, such a big animal!'"

Q: "What date did this take place?"

A: "1977."

Q: "Summer?"

A: "In the fall."

Author's note: Gary had his date wrong here, the event occurred in 1976, during the month of September. John Green quoted him earlier as saying it was September 1. If this is so, it was still summer.

Q: "Do you remember the month?"

A: "September."

Q: "Do you remember the day?"

A: "No."

Q: "Was it at night or day?"

A: "In the afternoon."

Q: "How much of this creature did you see?"

A: "I saw the whole thing."

Q: "For how long?"

A: "Oh...about five minutes."

Q: "Did it see you?"

A: "Yes. It got out of the river and took off."

Q: "How tall would you say it was?"

A: "About nine feet."

Q: "What color was it?"

A: "Black."

Q: "Did it have large arms?"

A: "Yeah, it had long arms."

Q: "Did it have a neck or did it seem the head was sitting on its shoulders?"

A: "The head seemed to just be sitting on its shoulders."

Q: "Did you see its face?"

A: "Yeah, it looked like, between a human and a dog."

Q: "A dog, did it have a snout or was the face flat?"

A: "It was flat."

Q: "Did it have dark skin?"

A: "I didn't notice that."

Q: "What distance were you from this creature?"

A: "About a block."

Q: "What's your friend's name, the fellow who was with you?"

A: "Peter Garyall."

Q: "Where is he now?"

A: "In Alabama."

Q: "American?"

Author's note: in his letter to John Green he said that his friend was a Greek immigrant. I suppose he was a Greek from the U.S.A.

A: "Yeah."

Q: "Did it make any threatening motions toward you?"

A: "No."

Q: "It just saw you and took off?"

A: "Yes."

Q: "Did you get a look at its feet?"

A: "No...I seen its arms, and its legs, it had long legs."

Q: "When it moved away, did it walk or run?"

A: "I would say it walked at a fast pace."

Q: "Did it ever go down on all four legs?"

A: "No, it stayed upright."

Q: "Did it make any noise?"

A: "No."

Q: "When you first saw it, was it kneeling in the water splashing itself?"

A: "Yes."

Author's note: In his letter to John Green he writes the creature was "standing" in the middle of the river. He makes no mention of the creature splashing itself.

Q: "And you watched it for how long?"

A: "About five minutes."

Q: "And it never made any noise?"

A: "No."

Q: "Did you notice if it was male or female?"

A: "I think it was male."

Q: "It had no mammary glands?"

A: "Not that I could see."

Q: "Did you shoot at it?"

A: "No."

Q: "Did it seem afraid of you?"

A: "No. It just seemed that it wanted nothing to do with us."

Q: "Did you report this to the police?"

A: "No."

Q: "Did you report this to anyone at the time?"

A: "No. I have told some people about it, but I don't think any of them believed me."

Author's note: Strange he didn't mention his letter to Cheam Publishing. I suppose he thought I was talking about immediately after the sighting. He didn't send the letter until about ten months later.

Q: "After it moved away, did you check for footprints?"

A: "Yes, but we didn't find any."

Q: "Did you smell anything?"

A: "No."

Q: "Do you think it could have been a bear?"

A: "No way."

Q: "You definitely feel you saw a sasquatch?"

A: "Yes."

There are some differences in his version of the story in this story in this interview when compared with the letter he sent John Green. I don't know if Gary ever knew that John did get the letter. However, most of the differences I have already pointed out. The similarities between the letter and the interview are impressive as well. Especially the description of the creature's face, between a human and a dog. I've never heard anybody use a dog before when describing the creature's face. Here now is a reprint of the letter he sent ten months after this incident took place.

Dear Sir or Madame:

I would appreciate it if you would give this letter to John Green or whoever is interested in the Sasquatch. On Sept. 1st of 1976 while I was hunting west of Caroline, Alta. With a Greek immigrant friend of mine, he broke through a clearing in the mountains. I shot a grouse in the bush, approximately a 100 yrds from my friend, when he called me to the clearing. A 1/4 mile away I noticed an animal standing in the middle of a fast flowing river. I said to myself, it had to be a bear or something. I started to run towards it, and I don't know why, but this animal waded out of the river, stood on the other side and looked at me and then headed into the bush. What I could plainly see was an ape about 8-9' tall, who had black hair, and a face between a dog and a man. When it walked its strides were in between 4 and 5 feet. I got up to my side of the riverbank and could not cross because the water was too fast, and the river was too wide. Me and the Greek are both plumbers, and we clearly saw an ape.

— Gary Schmidt.

When I left Gary's home he had difficulty walking me to the door due to his injuries. As I was getting into my vehicle he said, "Mr. Steenburg, you know now I wish I never saw that thing. Nobody believes me, and so I just don't talk about it anymore." I think most of the discrepancies between what he told me and what he wrote can be explained away due to the passing of time.

Another sighting was reported not too far from where Gary Schmidt claimed to have watched sasquatch in the Clear Water River. This incident occurred three years later during the summer of 1979. This report's location is on Route 591, which runs west from the town of Caroline, until it ends at the 940 Trunk Road. Just west of the bridge which crosses the Tay River lies the Tay River Campground. The Tay River flows into the Clear Water, just a few hundred yards to the south through thick trees. The Tay River Campground is on the north side of Route 591, on the west side of the bridge. The campground contains a few camping spots for RVs as well as cabins or cottages which are privately owned, some of which are leased during the summer.

Diane Menzel and her family were renting a cabin during the summer of 1979. One day she was walking along the road with her son. The two were approaching the bridge when Diane heard something behind her. She turned and was amazed to see a large hair-covered creature crossing Route 591, south to north. It was walking upright at a fast pace. Diane was quoted saying that the creature never did break into run, and it did turn its head to look at them. It

The late Professor Vladimir Markotic, stands on the spot where Diane Menzel and her son reportedly saw a sasquatch cross the road near the Tay River Campground, near Caroline, Alberta, during the summer of 1979. Vladimir who stood 5 feet, 11 inches tall, does not come close to the eight-foot creature which stood almost even with the road sign to the left. I took this photo from the spot where Diane and her son stood as they watched the creature cross the road, turn its head to look at them, then disappear into the trees to the left.

did not stop but just kept walking across the road, into the ditch on the north side, then up the hillside until it disappeared in the lodge-pole pine forest. Diane told the late Vladimir Markotic about the incident after a few months of silence, due to fear of ridicule. When she heard a forestry worker almost hit a sasquatch with his truck in the same area, she came forward with her story.

I have not talked to Diane herself, however I did receive a phone call from her son. His account matches hers to the letter with one exception. He made no mention of the creature turning its head to look at them as it walked. Some years later I was on a CBC talk show, *The Home Stretch*, discussing the sasquatch mystery. I don't think I even talked about this particular case on the show when a caller phoned in asserting that he was her ex-husband. He claimed to have seen the creature too that day, though he was farther down the road where the turn off to the rented cabins begins. From where he was, all he could add was that the creature was dark brown in color. I went to the sight with Vladimir in 1986. We took measurements from where Diane and her son were standing and the spot where the creature crossed the road. I found the two witnesses were approximately 240 feet from the creature when it crossed the road. If the positions of the two witnesses and the position of the creature are accurate, the creature was slightly uphill from them when it crossed the road.

In 1987, a Calgary TV news program did a three-part series on sasquatch encounters Alberta. In part two, Diane was one of two people on the program discussing their own encounters. Reporter Gary Bridgewater went over the story with her, and she again described the incident without any alterations from her original story. The forestry worker who reported almost hitting a sasquatch with his truck only a mile from the Tay River Campground has never been identified.

Another sighting near the Clear Water River took place in August, 1984. Al Lockman (not his real name) told me of seeing a twelve-foot creature while he was walking along a cutline near the Seven Mile Recreation Area. He agreed to come to my home for a full interview, however he didn't want his name to be revealed.

Q: "What date did this take place?"

A: "First week of August, 1984."

Q: "Was it at night or day?"

A: "It was at night at about 8:30. It was getting dark."

Q: "Describe the area in which this took place."

A: "Basically, I was walking down a cutline, it was heavy thick, dense bush on the left side, and I was basically heading due north. I was walking and it was really quiet out. It felt like something was watching me. I turned around, I was at a slow type of walk, slowing right down, and I heard something in the bush. I glanced to the left, you know to see what it was. I figured it might have been a range cow or something like that, but I couldn't see any other cattle around. I saw something that appeared to be about twelve feet tall.

"Roughly where I was standing it was thick heavy bush that come up to about eight feet, six to eight feet, and there was thick, heavy pine behind it, eh. And it was up against the hill a bit, so it was fairly dark. And I saw something that looked between a black and brownish color, just its head moving. I saw it move for three or maybe four feet possibly, then it disappeared off into the heavier bush. I could hear footsteps, about every three seconds. I would hear a loud thud through the dead fall, eh. Like I said it was strange because you couldn't hear anything at all. Like usually you can hear crickets and frogs at that time of night because there is some swampy area in there. It was so quiet you could almost hear your heart beat."

Q: "What distance would you estimate you were from the creature?"

A: "I would say, oh, forty feet maybe."

Q: "What was your first reaction?"

A: "When I saw it? Like I said, I stood there for about a minute and listened to it walk. And about every three seconds I would hear a loud thud as its foot hit the ground, and I knew whatever it was, it was watching me still, eh. So I hightailed it out of there. I walked down another cutline which was about fifty feet from the one I was on. They run parallel, so I just hightailed it down the second cutline to the campground. I didn't tell my wife about it that night in case she got worried and could not sleep. But I told her the next day that I saw something strange in the bush, and that was it."

Q: "What was it doing when you first saw it?"

A: "Gee, I couldn't really tell you, because all I saw was its head."

Q: "Did it stand and walk on two legs?"

A: "I would say yeah, because of the sound of it walking through the deadfall."

Q: "Did you see it go down on four legs?"

A: "No."

Q: "Was it covered in hair?"

A: "Its head was."

Q: "What color was it?"

A: "Brownish black."

Q: "How tall would you estimate this creature to have been?"

A: "I though it was about twelve feet."

Q: "What do you estimate its weight to have been?"

A: "By the sound of it walking, very heavy, at least 1,300 pounds...I think."

Q: "Did you see any facial features?"

A: "No, it was too dark."

Q: "Could you describe its arms?"

A: "No. That part of its body was behind the bushes. All I saw was its head."

Q: "Could you see if it was male or female?"

A: "No."

Q: "How long did you see this creature for?"

A: "Oh, I'd say that I saw it for maybe fifteen seconds as it walked toward the bigger trees. I would guess that I heard it for another minute as it walked away through the deadfall. It seemed to slow down as it walked into the thicker bush, because at first I could hear the thud of its footsteps every second. Then it seemed to be every three seconds, eh. I stayed there for about a minute, because there was a space of time, like about thirty seconds where I didn't hear it anymore. Then all of a sudden I heard it again. And that's when I decided to hightail it, because it didn't seem like it was going to leave and I really didn't want to hang around and find out what it was."

Q: "Did it ever make any noise?"

A: "No, it didn't make any noise at all."

Q: "Did it see you?"

A: "I would say yes. It had to, that's why it walked off into the bush. I think that it was watching me from the bush. It could see me, but I couldn't see it."

Q: "Did you smell any powerful odor before or during this sighting?"

A: "No."

Q: "After the creature moved off did you check for footprints?"

A: "No."

Q: "Did you report what you saw to park officials or police?"

A: "No."

Q: "Did you report it to anyone?"

A: "No. I mentioned it to a couple of guys once while having coffee at Montgomery and Denser, and they though I was crazy or something. I told my wife the next day, but never really told her that I figured it was a sasquatch or bigfoot. I just told her that I saw something weird in the bush, that I never caught a complete glimpse of it, but I didn't want to worry her."

Q: "In your own words, describe what happened."

A: "I was basically walking down the cutlines, looking ahead of myself, just going for a walk, get some exercise, stretch my legs, and I was maybe five blocks from the campground where we were camped. I was walking north along the cutline and...like I said I turned around, I heard something like a crack in the bush. I was starting to look around, it was fairly dark out. I had to focus for maybe five or ten minutes. It was getting dusk and I seen a head in the trees and then it started to walk off toward the heavier bush. It seemed like it walked maybe fifteen or twenty feet into the thick bush, then it stopped. I stopped and listened and it didn't appear to be moving anymore and it only went in maybe fifteen or twenty feet. So I figured, whatever it was it wasn't going to leave, and I didn't really want to stick around to find out what it was, eh. And as soon as I started walking again toward the other cutline, it started moving again. You could hear it. It wasn't moving as fast, but it was moving. I was pretty scared so I hightailed it out of there. I didn't run because I was afraid it might come after me, so I walked at quick pace, checking over my shoulder."

It almost sounds like the creature was intentionally trying to intimidate Mr. Lockman into leaving the area. Loud footfalls are not often reported. Could this have been a sign of distress or hostility? Or maybe the whole story is fiction and the loud footfalls are the sign of not doing your homework before making up a false report. After spending an afternoon with Mr. Lockman, knowing that he isn't really sure that this animal was a sasquatch, also requesting that I not reveal his true identity to anybody other than other researchers, I tend to think the former to be true. If it was a sasquatch, this was the last report, at least at the time of writing, of an extremely tall sasquatch in Alberta—August, 1984. I have not had any since.

Footprints have been reported on several occasions in Alberta. Some of which I've already talked about. One thing that is wrong with footprint reports, is that in most cases an individual who finds them does not photograph them. So we are left with just the individual's word that he found sasquatch prints. I'm always puzzled by the fact that today people will assume that almost any shapeless depression in the ground might be a sasquatch footprint. I have already in this book talked about footprint findings that upon examination turned out to be something else. The people were sincere in their belief that they had found something important. They were simply mistaken. When somebody reports finding footprints without photographs, we are left with only a description of the tracks to go on, which in my opinion is not good enough. I always have to wonder if the individual just thinks they are sasquatch footprints, and if I had the chance either to see them myself or look at photos it would turn out to be common animal tracks massed together or human footprints. It could also turn out that they indeed found good footprints, which makes the lack of photographs even more unfortunate. Of course we also have to be on guard against the hoaxer—something I haven't had to deal with in Alberta too often.

In 1987, I was sent a newspaper clipping from the *Edmonton Journal*, which was dated December 2nd, 1987. The article was sent by a Mr. Lid Orlando, who himself claims to have seen a sasquatch in 1954. I will be covering his sighting later in this chapter. He came across this article, and thought I would be interested. He was right of course.

Bigfoot Tracked

Calgary (special) - Lucien Lacerte says he isn't afraid if other people laugh at him a little. Almost 10 years ago he spotted two human like but hair-covered creatures running into the brush beside the Yellowhead west of Jasper. An now the legendary sasquatch has entered his life again. The weekend the Calgary ironworker returned to photograph large tracks he's convinced were made by a Sasquatch. Lacerte, 48, and a hunting partner found the prints in the snow Friday while hunting in the hills 32 km west of Claresholm. "About an inch of snow fell since then and partly filled in the tracks," Lacerte said last night, noting his Polaroid photos show little detail as a result.

"I don't care if people believe me or not. I took the shots for my own satisfaction." The tracks showed a stride of about four feet. Lacerte, who's hunted since his boyhood in southern Saskatchewan, said the creature's trails led directly over a two-foot high tree stump at one point. He said his partner, Larry McGillis, 35, saw the creature's tracks moving easily over thick deadfall that a man would have to clamber over. Lacerte added the two Sasquatches he saw while driving the Yellowhead ten years ago were black-haired and ran upright on two legs, with a very distinctive stride. "Sure, people smile sometimes, but I know what I saw, and those sure weren't bears."

I have been unable to locate Mr. Lacerte, or see the photos he took. The name of the reporter was not given and the *Edmonton Journal* didn't seem to know who wrote the article. They did say they would find out and ask him to contact me. I was never contacted by anyone. The man who was with Lacerte, Larry McGillis, has also proven hard to track down. Both men either have unlisted phone numbers or they had given the reporter assumed names.

A Mr. Stan Fisher, from Nanton, Alberta reported that he found a ten-and-a-half-inch footprint along a creek crossing, located about thirty miles north of Lundbreck, Alberta. He reported that the lone footprint sank three inches in damp sand where his own prints hardly marked. He did not photograph the print, but he did make a plaster cast of it. The size of this print is large but not beyond some human feet. Perhaps he came across a large barefoot human track left by somebody who was wading in the creek. Or perhaps it was a sasquatch track? I have not seen the plaster cast he was said to have made. This happened during the spring of 1972.

In 1985, a Mr. Kerr Patterson phoned me to tell of strange tracks he found a year earlier. He was working on a well in a new housing development west of Claresholm during the deer season of 1984. He could not recall the exact date. He had finished work for the day and he was walking a bulldozer out of the area. It was dark by this time, and he saw the tracks in the dozer's headlights. He told me they were humanlike (five toes), but large. He couldn't understand why somebody would be walking about barefoot in the country in deep snow. He took some of his friends to show them what he found, but they were not impressed. Mr. Patterson did not take any measurements of the prints, but said they were larger than any foot he had ever seen. The stride was about five feet. (Again I'm not sure if he really meant step, rather than stride.) The snow in the area was about a foot deep at the time. No photographs were taken. He also told me he didn't give the tracks much thought until he saw an article in the local press about two hunters seeing a sasquatch in the same area. I do not know any details about the sighting made by these two hunters, and I have not seen this article.

I was contacted by a Mr. Mike Doyichak of Brooks, Alberta, who told me about a strange creature he and Mr. Don Hankle saw along the 940 Trunk Road as the two were driving toward Coleman, after an unsuccessful day's hunting. They were about to enter the outskirts of Coleman when they noticed something standing on the side of the road. The two men slowed down trying to decide what it was they were looking at. The creature crossed the road left to right (east to west) and walked over a four-foot fence and proceeded up a hillside toward some trees. They followed the creature with the truck's headlights.

Before the creature was beyond the range of the lights, Mike jumped out of the car and yelled at the creature. "You up there, you better stop or we will start shooting!" The men had their rifles with them but did not take aim. "I really had no intention of shooting, I just wanted him to stop, if it was a man in a suit or something." The creature did not stop but continued to walk away until they lost sight of it. Mike told me that the creature was covered with black hair and that it was over six feet tall. At no time did it drop down on four legs, but covered the distance upright on two legs. The next morning the men went back to where they had seen the animal to look for hairs and footprints. None were found. There was no snow on the ground yet. This incident occurred on a Sunday afternoon in early November, 1987, at about 12:30 P.M. Mike declined my request for a full interview. Mr. Hawkle also declined.

Mr. Lidio Orland, who sent me the article from the *Edmonton Journal* entitled, "Bigfoot Tracked," earlier told me of his encounter with a sasquatch while he was hiking with his dog in the woods in the Crow's Nest Pass between the towns of Coleman and Blairmore, Alberta, during the late summer of 1954. Later he wrote me a letter with the details of his sighting.

> It was in the late summer of 1954. The place, about halfway between Coleman and Blairmore. About half a mile or less north of Highway 3. This is open hill country with mixed spruce, pine, and some poplar. The Highwood range to the east, and the Rockies to the west. I was out hunting rabbits, with a 22 cal

Winchester and my dog, "King". About 4 o'clock in the afternoon, I was sitting on a high rock ridge, looking east, when I spotted a tall black creature walking swiftly towards me. I just sat there watching, thinking it was man at first. But the closer it came the more unmanlike it became. Walking erect it covered the ground at an unbelievable rate. No human could possibly walk that fast, let alone run. When I first seen the creature, it was approx 400 yards away, and it was walking directly towards me to a point about 180 yards east of me. Then it turned abruptly right and headed N.W. up a ravine until I lost sight of it.

I don't think it saw me till just before it turned north. It walked directly under a large tree, with a large overhanging branch about 16 feet off the ground.

The Sasquatch cleared it by about 2 or 3 feet at the most.

The creature or Bigfoot was all black, but a lot slimmer than the photograph Roger Patterson took in 1967. It looked very much like a male. Again the most striking thing I can remember about the Bigfoot is the size, and the gait which was unbelievably fast.

This is the only report I've received of a sasquatch about twelve feet tall in the south of the province. All the others have occurred within 100 miles of the Big Horn Dam, near Nordegg. I do not know how his dog reacted when the creature approached, assuming King even saw the creature.

The Morley Indian Reserve, west of Calgary has had several reports over the years. The Stoney people have legends of sasquatchlike creatures using the Old Fort Creek as a route to the Bow River. Like Salish legends on the West Coast it is believed that seeing a sasquatch is an omen of coming tragedy. I have heard second-hand of a reddish brown creature seen near the town of Morley. One report came by in an unusual manner. I was discussing my second book on the *Terry Moore Show*, on QR77 radio in Calgary. One week later the show was talking about UFOs and other mysteries, when a man who said he was an Australian tourist phoned in to say he had just seen what he described as a reddish brown orangutan, cross Highway 1A, in front of his rented car. Terry Moore, forgetting I suppose that I was on his program only a week earlier, didn't realize that his man may be describing a sasquatch. He told the caller that the was unaware of the Calgary Zoo losing one of its orangutans, but he would check, and then he cut the caller off. When I was told of this I phoned in and suggested, off the air, that maybe this man had seen a sasquatch. Terry, remembering me, asked on the air for the caller to phone again. He didn't.

Willis and Elana Fox, who live on the Morley Reserve called me to report a sasquatch crossing the road in front of their car at the Old Fort Creek bridge. The bridge is an old single lane on Highway 1A. Oncoming traffic has to stop and make sure traffic is clear before proceeding. This old bridge has been replaced recently with a modern two-lane bridge, and a new lodge (Nakoda Lodge) has been built on the east side of the bridge. In May, 1983, when this sighting took place, construction had not yet begun.

Willis and Elana were driving east, approaching the bridge when they both saw a large, dark brown, hair-covered creature walk out onto the road just beyond the east end of the bridge. The creature hesitated for a moment, and looked at the oncoming car, then at a fast walk crossed the road, stepped over a four-foot fence in the ravine and disappeared in the large lodgepole pines. As the creature walked into the trees, the car had crossed the bridge and Willis had brought it to a stop with the intention of trying to get another look at it. He could see the branches moving as the creature disappeared deeper into the cover of the forest. Elana, realizing that her husband was about to leap from the car, started to cry, and she begged him to keep driving and not to go after it. Willis, concerned over his wife's panic in her pregnant state, did as she asked. I talked with both Willis and Elana, separately over the phone. They both said that the creature was covered with dark brown hair. The face was partly hairless, and it had black skin. Willis said the eyes were large and deep set within the forehead, and that the forehead jutted out somewhat. Elana said she did not really notice any facial features, except the black skin. Willis claims the creature stood between seven and eight feet tall, while Elana thought it was just over six feet tall. Both described its arms as big and powerful, and longer than a man's. Both said that the creature was very fast and stepped over a four-foot fence with no effort at all. Both said that the sighting was

Above: The bridge over the Oldfort Creek, on the Morley Reserve, where Willis and Elana Fox encountered a sasquatch in May, 1983.

Below: The four-foot fence that the sasquatch cleared with no effort at all, according to Willis and Elana Fox.

Photos: T. Steenburg, 1985.

between five and ten seconds in duration, and the thing walked upright on two legs the entire time it was in sight.

There was another sighting of a sasquatch on Highway 1A eight years later, about four miles east of the bridge. Wilford Fox (no relation to Willis and Elana in the previous story) of the Stoney Nation phoned me to report a strange animal he and a friend, Mr. Bill Ear, saw while they were driving west along Highway 1A, at approximately 11 P.M. on the night of May 23, 1991 (Thursday).

The two men were just talking when they noticed in the car headlights a large, hair-covered animal walking along the incline on the right-hand side of the highway. Wilford told me that the sasquatch was walking on two legs, and that it was traveling in the same direction as they were (west). When they slowed the car down the creature turned its upper body and looked at them while still walking. "It turned its whole upper body to look at us, not just its head and shoulders like a man would do!" He told me that the sasquatch was about eight feet tall, walked upright, and was covered with reddish brown hair. Despite the fact the creature turned and

The spot along the 1A Highway where Wilford Fox and Bill Ear encountered a sasquatch on May 23, 1991. After the two men passed the creatures, they stopped their car to look back and reported that a tribal police car came into view and put a spotlight on the creatures. However, the tribal police neither confirmed or denied that any of their officers saw a sasquatch that night.

Photo: T. Steenburg, 1991.

looked at them, he did not get a good look at its face. However, he did notice that the face was flat, and the skin was nearly black in color. The two men were stunned and drove past the creature which continued to walk westward along the highway. It didn't show any sign of fright or aggression, as the car passed no more than twenty feet from its left side. About 300 yards up the road, Wilford pulled over and got out to look back. At this point a patrol car from the tribal police came into view behind the animal. It stopped, and the car's spotlight was turned on, illuminating the creature. The creature turned right off the road, stepped over the fence and walked into the trees, where it was not seen again. The following Saturday, Wilford showed me the area where this creature was seen. I felt since there was a small pond with soft ground there should be some tracks. Nothing was found. I also examined the fence for 100 yards in both directions, looking for tufts of hair. None was found. Wilford said he would be willing to do a full interview, however every time I have tried to arrange this he has not been available. The man who was with him, Bill Ear, does not wish to talk about the incident, though he did tell me it did happen. I also contacted the tribal police, who informed me that they were unaware of any of their officers seeing a sasquatch on the night of May 23. If any of them had, they did not report it. Even though I have no doubt about Wilford's honesty, I feel that footprints should have been left in that area where this creature was. There was plenty of soft ground around the area where it crossed the fence, and headed toward the trees. There is hard ground too, and I suppose its possible it missed the soft spots. However, I think that would be unlikely.

Mr. Roy Hord, a Korean War veteran with The Lords Strathcona Horse, Royal Canadians, contacted me to tell about a sasquatch he saw while he was out horse back riding, near Eagle Lake, Alberta, in June of 1982. I went to his home to do a full interview about the day in question.

Q: "State your full name please."

A: "Roy Douglas Hord."

Q: "Where did this incident occur?"

A: "About half a mile from Eagle Lake, up in the Ya Ha Tinda."

Author's note: The Ya Ha Tinda Range has recently been closed to the public due to environmental concerns. It was a favorite spot for hunting, camping and horse back riding. In order to ensure that the public stays away, the government no longer repairs the only dirt road leading into the area. It is quite impassible now.

Q: "What date did this take place?"

A: "In the middle of June, 1982."

Q: "Was it at night or day?"

A: "Around 2 o'clock in the afternoon."

Q: "Describe the area in which this took place."

A: "Flat country with a little crest of a hill ahead, which this animal or sasquatch, or whatever it was, was just behind there."

Q: "Was the area open or treed?"

A: "There was spruce trees on both sides of the road."

Q: "What distance would you estimate you were from the creature?"

A: "About 125 and 150 yards."

Q: "What was your first reaction?"

A: "I just stopped walking and looked at it."

Q: "What was it doing?"

A: "Standing there, facing me."

Q: "Did it stand and walk on two legs?"

A: "Yes, it was on two legs."

Q: "Did you ever see it go down on all fours?"

A: "No."

Q: "Was it covered in hair?"

A: "What I could see was covered in hair."

Q: "What color was it?"

A: "Dark brown."

Q: "How tall would you estimate this creature to have been?"

A: "Seven feet."

Q: "What would you estimate its weight to have been?"

A: "400, 450? It was heavy."

Q: "Did you see any facial features?"

A: "Well it sort of reminded me of a teddy bear, the face I mean."

Q: "What do you mean by a teddy bear?"

A: "Well the short neck, and the ears were sticking out of it."

Q: "You could see its ears?"

A: "Oh, yes."

Q: "Were they like ours?"

A: "Just similar to a teddy bear's ears."

Q: "They stuck out?"

A: "Yeah, a little."

Q: "Did you notice anything else, eyes, nose?"

A: "No, not really."

Q: "Could you describe its arms?"

A: "The arms went down almost to the knees. When I saw it, the arms where just hanging by its side."

Q: "Could you tell if it was male or female?"

A: "No."

Q: "How long did you see this creature for?"

A: "About thirty seconds."

Q: "Did it ever make any noise?"

A: "No."

Q: "Did it see you?"

A: "Oh yes."

Q: "What was its reaction?"

A: "It just stood there and looked at me and the horse. Then it made a left turn and walked off into the trees."

Q: "Did you smell any powerful odor, before or during this sighting?"

A: "No."

Q: "After the creature moved off, did you check for footprints?"

A: "I looked, but it was rocky ground."

Q: "No footprints?"

A: "I didn't see any, no."

Q: "Did you report what you saw to rangers or police?"

A: "No."

Q: "Did you report it to anybody?"

A: "I did a couple of months after."

Q: "To who?"

A: "Just friends."

Q: "In you own words describe what happened."

A: "Well, I was up at the lake there. I was horse back riding. I was walking back with the horse, and that's when I seen this animal just over the crest of the hill."

Q: "What did the horse do?"

A: "Nothing. It saw it, but it didn't panic or anything."

It is interesting that the horse didn't react with any fear. During most encounters between people on horseback and a sasquatch, the horse usually will rear and panic. Bob Gimlin can attest to this. Since the horse was rented from stables in the Ya Ha Tinda, stables which no longer exist, I have to wonder if the horse's lack of concern was due to it being used to the presence of these creatures.

On December 13, 1991, I received a call from a Mr. Dale Jones (not his real name), who told me about a creature he observed while he was flying a small private plane in the area west of Sundre, Alberta, the previous October. He was over the forest a few miles east of the 940 Trunk Road, when he noticed below on a game trail, a large animal walk out of the heavy bush, cross the trail and disappear into the trees on the other side. It was only in the open for about five seconds. As it crossed the trail, it looked up at his airplane. Mr. Jones started flying in circles over the trees where the creature had disappeared. He thought he might have caught some more glimpses of it as it walked between trees. However, he was not able to keep it in sight. It was slightly overcast that day, and there was no snow on the ground yet. He described it as walking on two legs, dark brown in color and it stood about ten feet tall. Mr. Jones is a police officer with the Calgary Police Department. He did not report what he saw to local RCMP.

In September, 1988, John Green received the following letter from a Mr. Donald B. Kenny.

> On the 7th of November in 1975, I saw a hairy creature I believe to be a sasquatch. It was 8:30 in the morning. I had gotten out of my car, as I had seen a good size bull moose standing broadside to me in a swamp, 400 yards north of me. I saw there was roughly 1 feet of snow on the ground. I had a clear view of the moose. I loaded my .303 rifle and fired, knocking the moose down. Then my trouble started. I had shells in the rifle clip, loaded one behind the other, jamming my clip. I unjammed it and put it into the rifle. At the same time I looked up at the moose and

chambered another round into the rifle. At that very moment the moose got to his feet and slowly trotted out of the swamp, crossing a wild game and range cow trail. To say the least I was astonished to see the moose get up and take off. At the moment the moose crossed game trail and was heading uphill to the north east, the sasquatch came out of the trees on the game trail from the northwest and was walking much like a human, with what seemed like long arms. He appeared to stand 7'4" to 7'6", and weigh roughly 400-500 pounds, and was a reddish brown color.

The sasquatch proceeded to follow the moose uphill in a northeast direction. The sasquatch was clearly visible to me, at 400 yards in broad daylight, for roughly 3 minutes. It was more interested in the moose than me. The sasquatch did not look at me as I looked in disbelief at him for roughly one minute. You will never know the thoughts going through my mind at this time. After the moose and sasquatch in pursuit disappeared up the slope of the hill I crossed the swamp as fast as I could in a foot and a half of snow. I did not back track the sasquatch back up the game and cattle trail to the northwest. As I reached the game trail the sasquatch tracks were visible, as were the bull moose. Believe it or not I did not fear the sasquatch. I was more concerned with finding the wounded moose. With this in mind I proceeded to track the moose for two to three hundred yards. My trail ran out there, as the ground was extremely wet and soft and deep like sponge rubber, due the absence of snow.

At this point I came to a clearing on top of the hill. The sun was shining into a small opening in the trees, a clearing. It was deathly quiet, not even the birds were singing, or squirrels calling. A strange feeling came over me, as if something was watching me. I slowly looked directly to the east and then slowly I looked to the south, seeing nothing each time. I proceeded to slowly look in the northerly direction. I had slowly rotated from right to left when movement caught my left eye. I stopped rotating my looking and visually focused on the movement. And there, standing looking at me (for 30-45 seconds) from 30 yards was the sasquatch. It was then I feared for my personal safety, so I flipped the safety catch off of my rifle and aimed for a opening between two 10-inch pine trees. By then the sasquatch was moving, I figured in a northeastern direction. I was wrong, from what I could tell by tracking it for 50 yards it was heading due north.

Author's note: He did fire his rifle at this point, he continues.

After this I made three circles in a wider arc each time, to conceivably pick the tracks of the sasquatch or bull moose. I found no evidence of having wounded or killed the sasquatch. The sasquatch I believe was a male. The day the event took place could be between the 6th and 10th of November, 1975. The location was 8 miles north and 14 miles west of Sundre, Alberta.

It was thirteen years later, in September, 1988, that he wrote to John Green about this event. John forwarded the letter to me suggesting I follow up on it. This I intended to do, however I was in no hurry, seeing that this event had occurred thirteen years earlier. Later that same month I was a guest on another radio show. As usual, the host of the program gave my phone number over the air so anybody who believed they had seen a sasquatch could call. The next day Donald Kenny phoned me.

"I took a shot at a sasquatch in 1975," he told me. I thought this was a good chance to compare notes with what he had written to John Green. He agreed to come to my home for an interview. I gave no indication that I already knew of his encounter.

Q: "Please state your full name."

A: "Donald B. Kenny."

Q: "Where did this incident take place?"

A: "It took place ah, north and west of Sundre, in the year of 1975."

Q: "Was it at night or day?"

A: "Daytime."

Q: "What time of day was it?"

A: "I would say about fifteen minutes to nine in the morning."

Q: "Describe the area in which this took place."

A: "Ah, it took place directly north off a road, just off a swamp, which I estimate to be about 400 to 450 yards across."

Q: "What distance would you estimate you were from the creature when you saw it?"

A: "I'd say 430 yards when I first saw it."

Author's note: He wrote 400 yards in broad daylight, earlier. Close enough.

Q: "What was your first reaction?"

A: "One of surprise, ah I couldn't believe it, I swore a few words, you know?"

Q: "What was it doing?"

A: "It was walking on a cattle trail that the range cows normally use to travel from one area to another. It was in pursuit of a moose that I just minutes before had shot and knocked down, and the moose had got up and walked up this hill, and the sasquatch went after it."

Q: "Did it stand and walk on two legs?"

A: "Yes it did."

Q: "Did you ever see it go down on all fours?"

A: "No, I did not."

Q: "Was it covered in hair?"

A: "Reddish brown."

Q: "How tall would you estimate this creature to have been?"

A: "I'd say just under eight feet."

Author's note: Not as detailed in his height estimation as he wrote in his letter, however it is the same.

Q: "What would you estimate its weight to have been?"

A: "I'd say about 425 to 430 pounds."

Author's note: In his letter he stated that he thought the creature weighed 400 to 500 pounds. Again close enough.

Q: "Did you see any facial features?"

A: "I did, yes."

Q: "Could you describe them?"

A: "Ah, they were like...they're humanoid in nature, but...add the fact there is a lot of hair around the eyes and on the chin, and when it walked, it walked like the arms were below the waist. When it walked it walked with sort of a shuffle."

Q: "Did you notice the eye color of it?"

A: "No, I did not."

Q: "What about the nose?"

A: "It was not a prominent nose, it was not oversized, it was like a human's. It was not flat, it was not long, it was not wide or anything like that. It looked normal."

Q: "Did you see its ears?"

A: "The ears were not visible, no."

Q: "How about the mouth and teeth?"

A: "It did not open its mouth, it did not bare its teeth. What it did was when it saw that I saw it, it turned and looked directly south at me. It looked at me and I looked at it, and then all of a sudden it turned away and took off up the hill after the moose."

Q: "Could you describe the arms?"

A: "They are longer than a human's are, from the elbow up to the shoulder area they were very muscular."

Q: "Could you tell if it was male or female?"

A: "I could not tell, no."

Author's note: In his letter he stated he thought the creature was a male. Though he did not state why he thought this.

Q: "How long did you see this creature for?"

A: "I would say from five to six minutes."

Author's note: In his letter he stated he saw the creature for about three minutes. He also stated one minute at another point. Whether or not he was talking about two different moments is unknown. Still close enough in my opinion.

Q: "Did it ever make any noise?"

A: "No, it did not. And I might add I saw it in plain view, there was no mistaking what I saw."

Q: "Did it see you?"

A: "Yes, it did."

Q: "What was its reaction?"

A: "It didn't act startled, it just stopped on the trail as it was walking toward where the moose went up the hill. It turned and looked at me, and I mean eye-to-eye contact for about two minutes."

Author's note: Almost exactly the same as the letter. The only difference being the sasquatch stopped and looked at him for one minute.

Q: "Did you smell any powerful odor before or during this sighting?"

A: "Later on when I circled the bush to find the moose, I'd circled three times, and on the third time as I was looking for the moose I had come into a little clearing. I looked to the east, I couldn't see anything. I looked to the south, I couldn't see anything, and I just started looking to my left, I could see out of the corner of my left eye, I saw something move. I would say at a distance of about thirty yards. I could smell something. It wasn't out of the ordinary. I've been in the bush for many years, so it wasn't nothing out of the ordinary."

Q: "Did you check for footprints?"

A: "I did, yes."

Q: "Did you find any?"

A: "I think I found some in the snow, but they were not very distinguished."

Author's note: Here is the most noteworthy difference between what he told me and what he wrote in his letter to John Green. You get the impression from the letter that the creature's tracks went right along behind that of the moose. He didn't say here that there were no tracks, he just says that they were not distinguished. How this could be so, in snow from a creature he saw walking only minutes before, I cannot figure it out. One would think he would find fresh clear footprints in great numbers. Maybe he was a little embarrassed he didn't pay more attention to them at the time. Or maybe most of the prints were in that spongy rubberlike swamp land he described, and they didn't hold their shape. He does say that the sasquatch was walking up over a hill after the moose. Unless for some reason there was no snow on the hill as well, I still feel in an encounter like this there should be some footprints around.

Q: "Did you report what you saw to rangers or police?"

A: "No, I did not report this to any authorities, because then as I do now, I don't hold much credulity with these people, because all they do is record it and there is not much importance to it. So I had no reason to get in touch with them."

Q: "In your own words describe what happened."

A: "It was roughly 8:30, quarter to 9:00. I had stopped my car along this forest road because I'd seen a bull moose in this swamp. It was almost right over on the north side of this swamp. I got out of my car, I loaded my rifle. I took several steps away from my car into the ditch. The moose was standing broadside to me, I aimed and I shot. The moose did fall down, as the moose fell down and its four legs were dangling in the air, I attempted to load my rifle again, reload it. I had trouble with it, because I jammed it. As I was trying to unjam my rifle I looked up and this moose was getting back up on its feet, it proceeded to move rapidly away from me, going uphill onto a cow trail.

"At the point where this moose got onto the cow trail, that's when I saw the sasquatch come onto the trail. It walked about thirty-five yards and it stopped when it saw me. This lasted, as I said earlier, for a few minutes (looking at each other) then it stopped looking at me and it went after the moose. At this point after they both went out of sight. I went down to the spot where the moose had fallen to check for blood because, believe it or not, my primary concern was to get the moose. When I checked and found only a little bit of blood, I figured I'd better go after the moose and see if I couldn't get to it, so I went up the hill in the same direction that the moose and the sasquatch had gone. I had gone about 200 yards or so, I couldn't see tracks because we were above the area where there was snow and there was spongy moss on the ground so I couldn't track the moose very well."

Author's note: I guess he just confirmed here why he didn't see any good footprints. It does indeed seem that the hill was bare. He continues:

"So I made a wide circle and once I completed that I then make a larger one. I had just finished the second circle, and was about to start a third when...as I said earlier, I came into a small clearing. The sun was coming in from the east side. I looked east, I could not see any tracks. I looked to the south and I could not see anything that way either, and for some unknown reason because I've hunted for many years in the bush, I just knew that I was being watched. As I slowly rotated my head to the left I again made eye to eye contact with this sasquatch. This time I was only about ninety feet from it, and it was at that point I began to fear for my own safety, so I took the safety off my rifle, and as I started to bring my rifle around the sasquatch took off. I thought it was going to go in a northeasterly direction, so I aimed for a spot between two pine trees, and when I though the sasquatch was about to go by those trees I fired. After this I made a circle to see if I could see it again because I did not know if I hit it or not. I did not see it again."

At first I thought Mr. Kenny had made a change in distance when he said the creature was ninety feet away from him the second time he saw it, just before he raised his rifle. In his earlier letter he wrote that it was thirty yards away this second time. He is bang on, thirty yards is ninety feet.

Besides some slight differences, this interview is his letter only a little more detailed. I'm still bothered by the lack of footprints, however I think he explained the reason for this as well, but he does say at one point in his earlier letter, "As I reached the game trail, the sasquatch tracks were visible, as were the bull moose." He doesn't say that they were clear and sharp. In my interview with him he does say that tracks were there, though they were not distinguished. This confusion on my part may simply be caused by his slight lack of detail when he mentions the tracks in his letter. He didn't comment on the clarity of the prints at all. Also he was not carrying a camera with him that day so the footprints were not photographed, let alone the creature itself. Donald B. Kenny is an intelligent, well-spoken individual who really believes he saw a sasquatch that day. I only wish that he had brought the creature down that day with a well-aimed shot. If he had this ongoing question would have ended on November 7, 1975.

Recently I was visited by Mr. Tom Messer, of Calgary, who told me about a strange creature, he and four other people saw while they were four-wheel driving in the Waiparous/Meadow Creek area of Alberta, during the summer of 1981. Tom was driving, when he noticed the signs on my vehicle as I drove by. He phoned me later that evening. When he came to my home for an interview, he brought with him a friend, who produces videos, and has a keen interest in this subject. He wanted to tape the interview. I had no objections and told him to go ahead. George Tutt set his camera up, and taped Tom as he told me about the creature he saw sixteen years before.

Q: "Would you state your full name please?"
A: "Thomas Stefen Messer."
Q: "Where did this incident occur?"
A: "Near Waiparous, western Alberta."
Q: "What date did this take place?"

A: "I can't remember the exact date. The summer of 1981."

Q: "Was it at night or day?"

A: "Daytime."

Q: "What time of day was it?"

A: "Sometime around noon."

Q: "Describe the area for me in which this place."

A: "The area was wooded. We were on a cutline, and we just crossed over a range. One range, we just came down into a valley, in between two small, foothill ranges. It's all wooded, but in the bottom of the valley, was an open meadow. Quite flat, natural meadow, and we were on a kind of a cutline road, kinda winding around a little bit at the bottom. We came around...I was in the back...there were three trucks, and I was in the back truck, and we got into the bottom of the valley, you know, just crawling along, really slow. On the left had side, about maybe 100 yards into the meadow, to my left...there were some trees that came out from the circle of the meadow, and they came out into like a point. They extend out into the meadow a fair ways, and they end in the open meadow. What I saw was because I could see through the end of those trees, you know, the last twenty-five yards of those trees, I could see light through them, so the sasquatch (which I didn't know what it was at that time), but I just saw movement between those trees, I saw like a flash, then it went behind a tree, then it stopped, then another flash, and that is what caught my attention, like just the profile. I just saw something moving. So right away (we had C.B. radios), I just radioed the other trucks, I just said, 'There's something moving off to the left,' because I didn't know what it was at the time, and we all just came to a stop. At about the same time, the sasquatch came out at the end of the trees. Right at the end of that point, and just...it didn't run, it was a loose kind of a jog or something, whatever, it wasn't running, it wasn't walking. But it went straight across, didn't appear to be frightened, or spooked. It just appeared like it had somewhere to go, and away it went. But what I noticed was his speed. He didn't appear to be putting forth a lot of effort, he was going very, very fast, like he crossed that meadow, like...a heart beat. We never had a chance...ax, this one guy, he had a camera. He never had a chance to put the lens on it, because...he was out there, he went straight across and into the bush, and he was gone."

Q: "Besides yourself, how many witnesses were there?"

A: "There was four others."

Q: "Do you know their names?"

A: "Ah, some of them, I can remember their names, yeah. There was one person who was just a visitor. I can't remember the name. There was my wife, who's now passed away, she was there...there was Max Caruso, there was Harry Borro, and one other person, who came with Harry...I can't remember, it was the only time I ever met him. Harry Borro was my old foreman, at the job I used to work at. He's probably still around here. He's a Calgary native. Max I think is still here, don't see him that often anymore, ran into him a little while ago, but he's probably still around."

Q: "The reason you were out there, was it a pleasure trip, or were you working or what?"

A: "We were four-wheel drive enthusiasts. We were in the Stampeder Four Wheel Drive Club, and we used to go four-wheel driving all the time. So we were out there four-wheel driving."

Q: "What distance would you estimate you were from this thing when you saw it?"

A: "I would say 100 to 200 yards, somewhere between there."

Q: "What was your first impression?"

A: "My first impression was there was just something moving. It was just a movement through the trees. I saw it stop and go. My first impression was, 'What was that,' Ha. Ha. What is that, we got back together, cause it was only a second or two it was there...you know, we kind of looked at each other, like, did you guys see what I saw? You know, that type of thing. Am I seeing things? We just stopped and we talked about it for a while. We all agreed that it was a sasquatch, then we talked about what else it could be. We tried to think of anything else it could be. Could it be a bear? No, because it was going too fast up on two legs. A bear can go on two legs, but they can't go that fast up on two legs. Probably can't go that fast on four legs. We even thought, could it be a guy on a dirt bike, wearing black, you know? But there was no noise. We thought about everything else it could be, but there wasn't anything else."

Q: "How did you react, when you saw it?"

A: "I was just awe inspired. I was just blown away."

Q: "What was it doing?"

A: "It was just walking or jogging or whatever. Straight across the meadow. It didn't turn to look at us, it didn't seem to realize, or make any indication that it even knew that we were there at all. Maybe it did, maybe it didn't, but it didn't do anything to indicate that."

Q: "Did it stand and walk upright?"

A: "Upright, it walked upright."

Q: "Did you ever see it go down on all fours?"

A: "Never."

Q: "Was it hairy?"

A: "Yes."

Q: "What color was it?"

A: "Black."

Q: "How tall would you estimate this thing to have been?"

A: "I would estimate eight feet-plus."

Q: "How heavy do you think it was?"

A: "Oh man, five to seven hundred pounds. I would think five to seven hundred. It was a big animal, it was huge. I don't know, maybe even 1,000 pounds. I don't really know how to guess that."

Q: "Generally, you think it was big?"

A: "It was big. It was a long ways away, but it was still like...big."

Q: "Did you see any facial features?"

A: "Not distinctly. No."

Q: "When you saw this thing, did you see it from front, back, or profile?"

A: "Profile."

Q: "Could you describe its arms?"

A: "Long, and a big swing. It was different than other...bipeds? It was sort of the same kind of thing, but different somehow. It somehow managed to get the speed. You know, because of the way it moved perhaps. It didn't look normal, it didn't look like it fit with the geometric, or the mechanics of the bones or stuff like that, it didn't seem to correspond with how fast it was going...somehow."

Q: "Could you tell if it was male or female?"

A: "No, not for sure."

Q: "How long did you see this thing for?"

A: "Probably from start to finish...five to seven seconds, maybe."

Q: "Did it see you?"

A: "Not that I know of, but it may have."

Q: "Did it make any noise?"

A: "We couldn't hear it from where we were. It didn't make any verbal noises and we couldn't hear any bush noises because we were in the trucks and stuff."

Q: "Did you smell any odor before, during or after this sighting?"

A: "No."

Q: "Are there any other physical characteristics of this creature that stand out in your mind?"

A: "The size is awesome, the speed is awesome...it's just something that's very, very different that you...you just know when you see it, it's something different. I didn't get the chance to smell it, and look at prints and stuff like that, but..."

Q: "Did you check for footprints later?"

A: "No, we didn't. We were basically scared to go over there."

Q: "Did you report what you saw?"

A: "Not till now."

Q: "In your own words describe what happened, tell me the story again."

A: "Okay, we were in four-wheel drives, just trekking through the back country. We were in the Waiparous-Meadow Creek area. West of the forestry trunk road (940)...we actually camped over the first ridge, you go the first range back from the trunk road. We were camped in the meadow there over night. The next morning we got up and we started moving north, first of all we moved north. We came across a carcass, it was a cow, bull, whatever carcass, it was lying there rotting. We checked that out for a little while then we moved further north. We went up a...we turned west again when we got to a cutline, and we started heading west. We went up over another range on the cutline and came down onto the next valley. It was just as we were coming down into the valley, and the bottom meadow opened up, I got a glimpse out of the left eye, I saw this

216

dot, dash, dot, dash, through these trees, which came out to a point in the meadow. I radioed...at that time I just quickly radioed the other turcks that there was just something. I didn't know what it was, I hadn't identified it. Something moving to our left. Everybody stopped, we weren't going fast anyways, we were just crawling, and everybody stopped and we all looked, and that's when the...sasquatch, he just cleared the last three trees, and came out into the meadow, just went straight. I guess it was, but he was just moving. He didn't stop or make any noise, he didn't stop at all. He was going the same speed from one end to the other. He just went into the trees and that was it. That's what we saw."

Q: "Do you think it might have been a man you saw?"

A: "No. No way. No way. Ben Johnson can't run that fast."

Q: "Do you think that it might have been any other common animal?"

A: "No way. I've never seen any animal that fits this description. Standing on two feet, that large, no other animal. None."

I have the feeling that this creature had seen them, or at least the trucks. It was probably the sound of the approaching trucks which disturbed the creature in the first place, and when they came into view it decided to beat a hasty retreat. The result being five more people seeing the unbelievable for a few seconds before it disappears into the forest. I think Mr. Messer has been fascinated by what he saw for the last sixteen years. He told me later that he has gone camping at times hoping to see such a creature again. He has asked if he can accompany me on one of my trips in the future. He has, with George Tutt, started a small video production company in the last couple of years. The two have a keen interest in this creature now and hope to get video footage of one some day. Always taking their cameras along anytime they are in the woods west of Calgary, just in case. Tom regrets that his friend Max Caruso wasn't quicker with his camera that day. I found no reason to doubt his claim that he saw a sasquatch during the summer of 1981.

Bragg Creek, Alberta, is a wonderful mountain community, just thirty minutes west of Calgary. Many of the town's residents commute to the city to work, returning to Rocky Mountain wilderness each day. I almost moved there myself in 1985. The popular CBC television show *North of 60* was filmed there. However reports of sasquatch in the area have been few. One case did come to my attention however. Mr. Carl Melnyk, told me of a strange creature he saw while he was spending the day with friends in the Bragg Creek Provincial Campground in June, 1974. Mr. Melnyk is a hunter, and told me that he always thought if he ever saw such an animal, he would have seen it well out hunting, deep in the back woods. Not in a public campground near a town, and only thirty minutes from a major city.

Q: "Would you state your full name please?"

A: "Carl Melnyk"

Q: "Where did this incident occur?"

A: "It occurred west of Calgary, near Bragg Creek, in a provincial campground."

Q: "What date did this take place?"

A: "It was in 1974, the month of June."

Q: "Was it at night or day?"

A: "It was during the day."

Q: "What time of day was it?"

A: "It was about three or four o'clock in the afternoon."

Q: "Describe the area in which this took place."

A: "I was sitting on the riverbank, a raised riverbank, and it was down at the edge of the river. It was in some shrubbery by the tributary that lead into the Elbow (River)."

Q: "What distance would you estimate you were from this thing when you saw it?"

A: "I was about...100 feet."

Q: "What was your first reaction?"

A: "My first reaction was that I didn't know what it was that was going on. I just...watched, and I didn't have any fear. I was a little shocked, so I probably wasn't believing my eyes."

Q: "What was it doing?"

A: "It was just rambling around in the trees or the shrubbery."

Q: "Did it stand or walk on two legs?"

A: "It walked on two legs."

Q: "Did you ever see it go down on all fours?"

A: "No I didn't, no."

Q: "Was it covered in hair?"

A: "Yep."

Q: "What color was it?"

A: "It was brown or black. I guess in between, a dark color."

Q: "How tall would you estimate this thing to have been?"

A: "I thought it was huge. I had an overview. I was looking down on it, but I was thinking ten feet."

Q: "What would you estimate its weight to have been?"

A: "Oh, I would say 600 pounds, at least. That might be a bit much, since I'm not very good at estimating. However it depends because it was pretty muscular. It looked big, bulky, husky. That's why I think it's that much."

Q: "Did you see any facial features?"

A: "No. Nothing distinctive, because of the distance, and it was in the trees too. I didn't notice anything like that, just a figure on two legs."

Q: "Could you describe its arms?"

A: "Yeah, they were long, they were rather thin though, they weren't big arms, but they were long."

Q: "Could you see if it was male or female?"

A: "No."

Q: "How long did you see this for?"

A: "I think about fifteen or twenty seconds at the most."

Q: "Did it ever make any noise?"

A: "No noise, but I do think that it knew I was there though."

Q: "Why do you say that?"

A: "Just the fact it left in such a hurry."

Q: "Was it walking or running when it left?"

A: "It was a fast walk."

Q: "What direction was it going? Towards the town of Bragg Creek?"

A: "No, it went up river. Back towards the foothills."

Q: "Did you smell any powerful odor before or during this sighting?"

A: "No, none at all."

Q: "After the thing moved off, did you check for footprints?"

A: "Yes I did."

Q: "Did you find any?"

A: "No."

Q: "Did you report what you saw to the police?"

A: "No."

Q: "Did you report it to rangers?"

A: "No officials, just friends."

Q: "In your own words describe what happened."

A: "Well, we were just getting ready to cross the river, and I was sitting on the bank. I was looking down on the river, and I noticed to my left, which would be west, cause I was facing north. I noticed that there was a something or other moving around in the trees. At first I thought that it was an elk or a moose, because of the color. After a few seconds I noticed that it was moving on two legs, and that's when I realized that it was not a normal animal, at least not common. So I just stared at it for about fifteen seconds or so, and then it just took off in a big hurry. It was the speed that made me think that it was not a man."

Q: "You don't think that this could have been somebody in a costume?"

A: "I don't think so because it was moving way too fast, and I don't think it was moving as fast as it could either, and it went just ripping through the trees you know, like there was nothing there."

Q: "Were you the only one to see this thing?"

A: "Yes, I was the only one."

Mr. Melnyk later explained that he was sitting there on the bank waiting for his friends to catch up with him, before they would continue on their hike. When his friends did catch up with him, he told them about the creature he had seen. It caused great excitement among the group, and they all spent the rest of the afternoon looking for it again. However, no sign of the creature was found, resulting in some doubt among his friends that he was telling the truth. Some friendships ended that June day in 1974.

I will end this chapter with a report that was even closer to Calgary than the town of Bragg Creek. On August 1, 1991, I received a phone call from a Mrs. Lorna Vancam (not her real name). She told me that her son, Zachariah, had seen some sort of animal on his grandfather's hobby farm the day before. She told me that whatever it was her son saw, it shook him up pretty bad. I talked to her son on the phone and he briefly told me what he saw. Next I talked to his grandfather, who owned the property where the sighting took place. He agreed to show me the spot the next day, on the condition that I keep the precise location of the property a secret, and that I leave my rifle at home. The next day, Friday, August 2nd, two days after the sighting, Zachariah and his grandfather showed me about the property. It was sort of a family expedition with Zach, both his grandparents, his mother and his two brothers scanning the ground for footprints. Both the boy's mother and grandfather told me they questioned the boy continuously to determine if it could have been a bear or moose. But Zach insisted that the creature he saw walked upright like a man and did not go down on all fours. Before we all started to look around the property, I interviewed Zach (age eleven), with his grandfather looking on, about the creature he saw. I found Zach to be a straight-forward young man, not the type who likes to spin tall tales.

Q: "What date did this take place?"

A: "July 31st."

Q: "Last Wednesday?"

A: "Yep."

Q: "Was it at night or day?"

A: "It was during the day."

Q: "What time was it?"

A: "Between two o'clock and three thirty in the afternoon."

Q: "Describe the area in which this took place."

A: "It was in the trees, just off a little path."

Q: "What distance would you estimate you were from this thing when you saw it?"

A: "Fifty to a hundred feet."

Q: "What was your first reaction when you saw it?"

A: "I was scared to death, that was my first reaction."

Q: "What was this thing doing?"

A: "It was walking really fast, taking long strides."

Q: "Was it walking towards you or away from you?"

A: "It was walking up the hill, and then it turned away from me, and it kept on walking."

Q: "Did it stand and walk on two legs?"

A: "Yes."

Q: "Did you ever see it go down on all fours?"

A: "No."

Q: "Was it covered in hair?"

A: "Yes."

Q: "What color was it?"

A: "The hair seemed white on the bottom, and got darker towards the top."

Q: "How tall would you guess it was?"

A: "Oh...seven feet maybe."

Q: "How heavy would you say it was—heavy, light?"

A: "Heavy."

Q: "Heavier than a man, lighter than a man?"

A: "Heavier than a man."

Q: "Did you get a good look at its face?"

A: "No."

Q: "Could you describe its arms, shorter than a man's, longer than a man's?"

A: "Yeah, they came down to about the middle of the top part of the leg."

Q: "The arms, were they covered in hair as well?"

A: "Yes."

Q: "Could you see if it was male or female?"

A: "No."

Q: "How long would you say you saw this thing for?"

A: "Ten to fifteen seconds."

Q: "Did it make any noise?"

A: "Well, when it got up to the top of the hill, it would walk about two steps, then make a noise, like when you are tired after a race or something, it would take two steps and make this sound, a real deep sound."

Q: "Like a grunt?"

A: "Yeah."

Q: "Did it see you?"

A: "I think it might have. When I first saw it, it sort of turned its head away."

Q: "Did you smell anything?"

A: "No."

Q: "Did you see any footprints around after it was gone?"

A: "I never really searched after that."

Q: "Did you report what you saw to the police?"

A: "No."

Q: "Did you report it to anyone?"

A: "Yes...you."

Q: "In your own words, describe what happened."

A: "We had a bailer, and we were bailing some hay, and we needed a grease gun, so I walked up to the garage. And when I got inside of the garage I sort of thought something was near. I went back down the path about ten to twelve feet. I heard a noise and I saw it. It walked with its arms still, taking long strides. It walked up to our tree house, then it turned and walked in the opposite direction away from me. I just stood still and laid down on the ground."

Q: "Was it on the right side of the trail or the left?"

A: "To my right."

After the interview Zach showed me the area where his sighting took place. His grandfather was with us. I measured the distance from where he stood to where he thought the creature stood and walked. If accurate, it would be a distance of about eighty-five feet. We did find impressions in the tall grass that looked like they could have been made by the creature's footfalls. However, none were clear footprints. I did measure the step between these flattened footfalls. It measured eight feet, large even for a sasquatch. He said that the creature was walking at a fast pace, not running.

Later that day Lorna found what might have been a footprint on a vehicle path, about 200 hundred yards from where her son had seen the creature. The path was typical of a rarely used vehicle trail, tire marks, with tall grass in between. The print was not deep enough to cast, however toe impressions could be made out. The print was thirteen inches long, six inches across at the ball of the foot, and four inches across at the heel. Later as we searched the perimeter fence, we found a spot where something had snagged itself on the top strand of wire pulling it free from three fence posts. The grandfather told me deer often get caught in his fence and he often has to do fence repairs, patrolling his fence at least once a week. This area was undisturbed the day before Zach had his encounter. I found brown hairs stuck to the top strand of wire. I sent these hairs to Dr. W. Miller, of the Molecular Immunology Lab, Center for Biologies, Evaluation and Research, FDA, and to Dr. David Metzger of the Forensic Science Laboratory, Illinois State Police. Dr. Metzger was to determine what animal these brown hairs came form. If he could not identify them, Dr. Miller would do DNA analysis on them. After about three months I was informed that the hairs found on the fence were those of a bighorn sheep. Bighorn sheep are common only thirty minutes west of the property, however they have never seen one on the property.

Could Zach have seen a panicked bighorn sheep bounding away from him? Judging from his description, and the fact a footprint was found, I don't think this is possible. All this really means, as far as I'm concerned, is the creature Zach saw was not what snagged itself on the fence. Zachariah Vancam did see something that shook him up, and caused great concern from his family who questioned him extensively about what he saw. The creature was not seen again on the hobby farm. About two weeks later I heard second-hand that somebody reported seeing a white and brown sasquatch crossing Highway 22, northwest of Cochrane. The highway is only about five miles west of the hobby farm. Anything traveling west would have to cross this highway at some point. However, I have no more details on this incident.

I do have many other file reports from the province of Alberta, which I have investigated. However, I have written about most of these in my two previous books. When I started researching sasquatch in 1979, very little had been recorded concerning eyewitness encounters with this creature on the eastern side of the Rocky Mountains. As I said at the beginning of the chapter, there is no wall between B.C. and Alberta, so if sasquatch have been reported in eastern B.C., they must have been seen in western Alberta as well. These were my thoughts when I started my own investigations. I think I can safely say, now after all these years, I was right.

Above: The tree house which the creature walked up to before changing direction and walking out of sight.

Below: The fence on which something caught itself, leaving tufts of hair behind. It was thought maybe the sasquatch that Zach Vancam (not his real name) saw had snagged some hair on it. It later turned out to be bighorn sheep hair.

Photo: T. Steenburg, 1991.

Above: A possible lone footprint found by Zachariah's mother, a hundred yards or so from where the boy saw the creature. The print was 13 inches long. If you look close- ly, you can see the five toes on the left side of the print.

Below: The lone footprint measured 6 inches across at the ball and it measured 4 inches across at the heel. The print was not deep enough to make a plaster cast.

Photos: T. Steenburg, 1991.

The Crandell Campground Incident

This encounter between four witnesses and sasquatch also occurred in the province of Alberta. I've put it in a chapter of its own due to the fact in the twenty years I have been researching this mystery of the sasquatch, this is the most interesting and detailed account I have ever looked into. I feel in the future, assuming a sasquatch is not brought in, the Crandell Campground incident will be considered one of the sasquatch "classics." This fascinating story took place in the Crandell Campground, located in Waterton Lakes National Park, in the southwest corner of the province. There were two couples with two cars at one campsite. This incident occurred on the Victoria Day long weekend of 1988.

I first heard of this incident three months after it happened when Susan Ray Adams, contacted me and told me her story. She called me on the night of August 24, 1988, after seeing my ad in the Calgary press. I made arrangements to visit her home and interview all four witnesses separately. So impressed was I with this story, I delayed production of my first book *The Sasquatch in Alberta*, so some details could be included. I also put portions of the interviews in my second book *Sasquatch/Bigfoot: The Continuing Mystery*, which was published in 1993. Here I will include all four interviews in their entirety. To save space I will print the question, then give the answer from each of the witnesses in order. Remember the witnesses were not interviewed as a group, but apart, out of hearing range of each other.

When I visited Susan's home, all four questioned me extensively about sasquatch in general. It was obvious all were still on pins and needles three months after their encounter with a sasquatch. I was also handed a copy of a handwritten letter which one of the four, Darwin Gillies, wrote for the park warden's office the next day. I reproduce it here.

Monday, May 23.
At approximately 12:50 a.m. at the Crandell Lake Campground we spotted a very unusual animal. We were sitting

at our campfire when we heard some snorting. We assumed it was a deer, but upon further observation we decided it was a bear, and bolted for the cars. The animal was on its hind legs, and we switched on the headlights, on one of the vehicles. From the shadows, I could see the animal was moving on its hind legs so I called to the other vehicle to turn on its lights.

What we saw was incredible. This animal was not only on its hind legs, it was striding, like a human. We watched as it walked through the trees, for at least three to four seconds. Immediately thought it was a joke. We're all convinced it was not a bear. We jumped into the same vehicle and followed in the general direction it disappeared. We came across another vehicle and flashed our lights. These people had also sighted something very strange and were quite scared. This confirmed that we had both seen the same thing. It is important to note that we are four mature, responsible, professional people. We thought very carefully before coming in to report this incident at the warden's office. All four of us are convinced that it was not a bear. I am equally convinced it was not a practical joke. If it was, it was pretty elaborate and well done. From our sighting, the description we can give you is as follows:

The animal was approx 8 feet tall (as measured by the tree it was standing beside in our campsite). The animal was never on all fours. When we switched on the headlights and got a good look, this thing was striding, and big strides at that. It also had long arms which were swinging while it moved through the bush. It wasn't a bear O.K.! I don't know what more I can write about this incident. Would appreciate hearing anything that might explain what we saw, (or additional sighting, if any).

Darwin Gillies

Susan requested a copy of Darwin's written statement from the warden's office. Park Warden Alan Dibb replied with a short letter:

May 26/88

Susan:

Here's a copy of the statement, as you requested. Nothing else to report. Crandell Campground has been nearly empty all week. I'll let you know if we do hear anything.

Alan Dibb

Warden Service

The date on the letter show that Alan Dibb wrote to Susan on May 26, 1988. On July 10, Mike Sheen and three boys had their campsite destroyed by some animal which made high-pitched screaming noises (see previous chapter) by Twin Lakes, in Waterton Lakes National Park. I know that the wardens who investigated this incident knew about the events in Crandell Campground, the previous May. However Susan was not contacted or informed by the warden's office of this other incident.

As I said before I interviewed all four witnesses separately, but I will give the answers to my questions from all four, in order, to save space.

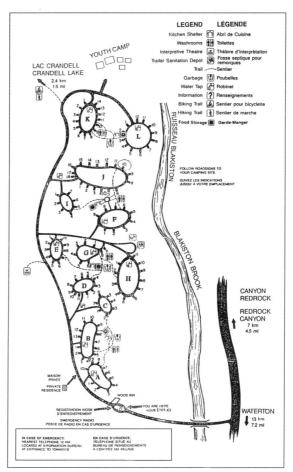

Crandell Campground on the Blakiston Brook in Waterton Lakes National Park. Site C-3 is where the events of May 23, 1988, occurred.

Q: "Would you state your full name please?"

Susan: "Susan Ray Adams."

Scott: "My name is Darell Scott Stoness, but I go by Scott."

Darwin: "Darwin J. Gillies."

Shannon: "Shannon Senkow."

Q: "Where did this incident occur?"

Susan: "In Waterton Lakes National Park."

Scott: "I was in Waterton Park, and I believe the name of the part we were in is Crandell."

Darwin: "Site C-3, Crandell Lake Campground, Waterton National Park."

Shannon: "In Waterton National Park."

Q: "What date did this take place?"

Susan: "The Sunday of the May 24 long weekend, 1988."

Scott: "It was the long weekend in May, on the Sunday. Actually it was Monday, because it was early into the morning."

Darwin: "May 23, 12:50 a.m."

Shannon: "The May long weekend."

Q: "Was it at night or day?"

Susan: "At night."

Scott: "It was at night."

Darwin: "At night. Or early morning."

Shannon: "It was at night."

Q: "What time of night was it?"

Susan: "Approximately, ten to one in the morning."

Scott: "About one in the morning."

Darwin: "12:50 a.m."

Shannon: "It was around one o'clock at night."

Q: "Describe the area in which this incident took place."

Susan: "It was in the outskirts of the campsite. The campsite is located about...fifteen miles outside of Waterton town."

Scott: "Typical area for mountains, it was rocky and trees, and ridges of rocks with moss on them"

Darwin: "Typical campground, circular gravel road, and we were on an elevated campsite, about three feet high off the main path."

Shannon: "The campsite itself is open. Ummm...a lot of trees."

Q: "Was it heavily treed?" (Susan only)

Susan: "It was treed. There was a stream, but the stream was about a quarter of a mile away."

Q: "What distance would you estimate you were from the creature?"

Susan: "When I first saw it, I would say...I would be within ten feet of it, I startled it."

Scott: "I saw it twice. The first time it would have been ten to fifteen feet. The second time, probably around forty-five feet."

Darwin: "When I initially saw the creature I was roughly twenty to thirty yards away."

Shannon: "Twenty feet maybe.

Author's note: Susan and Scott both saw the creature first as they walked along the trail to the washrooms. Darwin and Shannon didn't see it until the lights of the cars were turned on.

Q: "What was your first reaction?"

Susan: "My first reaction was that it was a bear."

Scott: "My first reaction was...I didn't believe it, but when I turned on the lights (car lights) it occurred to me that this could be a sasquatch."

Darwin: "My first reaction was surprise. I thought initially it was a practical joke, because I knew it was no bear."

Shannon: "I thought it was a bear, so I was scared and ran for the car."

Q: "What was it doing?"

Susan: "It was standing and looking at us, and then it grunted at us, as we approached it. (She is referring to the first sighting on the trail.)"

Scott: "The first time we saw it, me and my wife were going to the washrooms to brush our teeth. And it was standing right on the path watching us, and as we got closer to it, it snorted at us like a big bull. I don't know if you have heard how a bull snorts when he's chasing you, but I have. And that's what I thought it was, but I thought it was a deer, so I kept on pulling Susan. Finally she yelled, 'It's a bear!' and everybody ran to the cars."

Darwin: "It was standing there."

Shannon: "It was standing there."

Q: "Did it stand and walk on two legs?"

Susan: "It was standing on two legs."

Scott: "It was always on two legs."

Darwin: "It was walking on two legs when we saw it."

Shannon: "Yes."

Q: "Did you ever see it go down on all fours?"

Susan: "Never."

Scott: "No."

Darwin: "Never."

Shannon: "No, never."

Q: "Was it covered in hair?"

Susan: "Yes."

Scott: "Yeah. I believe it must have been dark brown or black, because I looked straight at it on the trail, and I didn't notice it until after I thought back about it."

Darwin: "Yes."

Shannon: "Yes."

Q: "What color was it?"

Susan: "I would say it was...black or dark brown."

Scott: (He answered this question already, so he was not asked again.)

Darwin: "Black."

Shannon: "I couldn't tell the color, but it was dark."

Q: "How tall would you estimate this creature to have been?"

Susan: "I would estimate it to be eight feet tall, as measured on a tree, because we checked the next day."

Scott: "Between seven and a half to eight feet."

Darwin: "Initially I thought it was quite tall. I think in my anxiety or what not. After talking with Scott we estimated about eight feet tall as measured on a tree it was standing beside."

Shannon: "Between seven and eight feet."

Q: "What would you estimate its weight to have been?"

Susan: "Oh gosh...it was pretty heavy. It was huge. I don't know, maybe 600, 700, 800. I have no idea."

Scott: "If it was proportionate to me? I'm six feet tall, so I would guess 500 pounds."

Darwin: "Three-hundred-plus pounds."

Shannon: "That's a good question. I would say very heavy. At least 300 pounds."

Q: "Did you see any facial features?"

Susan: "Not really, there was nothing pronounced, like a nose or anything, it was kind of a flat face I would say."

Scott: "No."

Darwin: "None at all."

Shannon: "No, none."

Author's note: I tried for more detail from Susan with this question.

Q: "It was too dark to see its eyes, nose, mouth?"

Susan: "Well, I saw it had eyes, nose and a mouth, but..I really didn't see anything distinctive about it. It was just a flat face I guess."

Q: "What color was its skin?"

Susan: "I don't know."

Q: "Could you describe its arms?"

Susan: "Yes. They were very long. The tips of its finger came very close to its knees. They were really long."

Scott: "Any I didn't notice its arms as much as its legs."

Q: "Legs?"

Scott: "The thing that was most remarkable to me about this thing, was that I watched it from the time I turned on my headlights, when we were looking strait at him, and he took maybe three strides to cover the distance that the headlights illuminated—thirty, forty feet. So I would say that he had about a five-foot stride, and he moved very quickly, and his legs were very straight as compared to a human when he walks. This animal kind of swung them straight with very little bending of the knee."

Q: "Was it walking or running?"

Scott: "Walking."

Darwin: "They were long arms. It was a lanky type creature, so the arms I would say, came well past the waist."

Shannon: "Long, very heavy, hairy."

Q: "Could you see if it was male or female?"

Susan: "No."

Scott: "No, but ah...if females have breasts? Then I would say that it was male because we didn't see any breasts."

Darwin: "Not at all."

Shannon: "No. I would guess that it was male."

Q: "How long did you see this creature for?"

Susan: "Total?"

Q: "Yes."

Susan: "A minute...two minutes maybe? Probably closer to a minute."

Scott: "I saw it twice. Once while we were heading down the trail, and I would say that was probably a total of ten seconds. Then I got back into the car, turned on the lights, and I think I saw it for a grand total of three or four strides, so that would probably be about another ten seconds."

Darwin: "About three or four seconds."

Shannon: "When we saw it in the car lights, it could have been no more than twenty seconds. Ten to twenty seconds."

Q: "Did it ever make any noise?"

Susan: "Yes, it grunted at us."

Scott: "Yeah, like I said before, it made a sound like a bull when he's chasing you, and I can remember that because I was chased by a bull once. So it stands out in my mind. I reflected on the sound later because I don't think a human being could imitate that sound because it so...like it was an animal with a big throat, and blowing a lot of air."

Q: "Was it a snort, scream, grunt?"

Scott: "A snort"

Darwin: "Initially when Scott and Susan left the campground and walked towards it, it made a loud...snort or grunting sound. When they thought it was a deer and continued to walk toward it, it grunted again."

Shannon: "Yes."

Q: "Could you describe it?"

Shannon: "It was some type of grunt."

Q: "Did you see it?"

Susan: "Yes."

Q: "What was its reaction?"

Susan: "It grunted at us. It warned us, because we were coming right at it."

Q: "Did it seem scared, hostile?"

Susan: "It didn't seem hostile. It didn't seem scared. It just seemed to say, Yeah I'm here, and it kind of grunted at us. Like a warning."

Scott: "Most definitely."

Q: "What do you think its reaction was?"

Scott: "Well, I would say that this animal was never scared. He was ten maybe fifteen feet away from us. Looking right at us, I would say more startled than anything, like we were coming down the trail right towards it, and all it did was snort."

Q: "When you were walking down the trail, it was standing there watching you?"

Scott: "Yes, looking straight at us."

Darwin: "Yes it did."

Q: "What was its reaction?"

Darwin: "I believe it was just curious. It stood near our campsite a distance...I'm convinced that it was watching us for quite a while at our campsite. The reason being, is when the cards blew off the table earlier, about fifteen minutes before this incident occurred. The cards blew off the table. I thought I heard this creature before make this sound, because one of us ran towards the bush to retrieve the cards. So I'm convinced he was standing there watching us with curiosity, and he was not afraid of us."

Shannon: "I'm sure it did, yes."

Q: "What was its reaction?"

Shannon: "Well I think that it had been watching us for awhile, and it just wandered off on its way."

Q: "Did you smell any powerful odor before or during this sighting?"

Susan: "No."

Scott: "No."

Darwin: "No. We were up wind."

Shannon: "No."

Q: "After the creature moved off, did you check for footprints?"

Susan: "The next morning we did, and we didn't find any."

Scott: "Yes, I did. But in that area, like I said before, it was rocky with a little bit of moss on top of it. So you could have had an elephant move through there and it wouldn't leave prints that I would be able to distinguish."

Q: "Did you check for footprints right after the sighting?"

Scott: "No, the next day."

Darwin: "The next morning we did. We didn't check that night. We weren't going anywhere near the bush."

Shannon: "The next day we did, but we didn't find anything."

Q: "Did you report what you saw to park officials or police?"

Susan: "Yes."

Q: "Did you report it to anyone else?"

Susan: "Apart from friends, we've told nobody except you."

Scott: "To the warden at Waterton National Park."

Q: "And what was his reaction?"

Scott: "I think he believed we saw something. He tried to tell us it must have been a bear. But over all I would say he was very objective, and if another sighting came up, he would want to know about it. If it was a fraud, he didn't want it scaring the campers away, but I would say it was about fifty-fifty in his mind that we saw what we said we saw."

Darwin: "Yes, we did to the park officials. Later on I told some people about it, and they...there are some pure skeptics out there, like I was before."

Shannon: "Yes, we talked to the warden the next day."

Q: "In your own words describe what happened."

Susan: "It was about ten to one in the morning on the Sunday as I indicate. We had been playing hearts, we had been out there a couple of days, so we were relaxing a bit. It was really windy and our cards got blown off the table, and I was getting quite tired. My husband and I decided that we would go and brush our teeth and then go to bed. We went to the car, got our toothbrushes and toothpaste, and we started down the beginning of the trail, that was very close to where our campfire was, and where Darwin and Shannon still were. My husband held me by the hand and we started down the trail. I thought I heard something and I told him, 'I think I hear something,' but he really didn't take notice of that, because I always hear stuff because I'm kind of scared at night. We took a couple of more steps and then we saw a big hairy creature standing up in front of us, probably ten feet away, and it grunted at us three times. Well, I screamed, 'It's a bear!' and I ran and got in the car. One of the other fellows, Darwin, he got into the car with me, while Shannon ran into the other car, and we looked out and Scott still wasn't in the car! He was backing up slowly, he checked my door, I wouldn't let him in, so he went to the other car Shannon was in."

Author's note: I always laugh when I recall how Susan locked her husband out of the car and wouldn't let him in. I guess somebody had to be sacrificed.

"I was quite upset. I wanted to honk the horn and wake up people in the campground, but Darwin put his arm on me and said, 'No, just calm down, relax,' and then he said, 'Turn on the headlights.' The two cars were bumper to bumper, so we turned our headlights on, and we looked around, I didn't see anything, and Darwin thought he saw something move by the fire, towards the trail, but I didn't see anything. Then he yelled to Scott in the other car to turn on his headlights. Scott turned on his headlights and we were looking out the back of the hatchback we were in. My husband in the other car turned on his headlights and in about ten seconds after the headlights were turned on, the creature walked into the light.

"It was walking on a ridge about...I'm bad at guessing distance, but I would say about thirty or thirty-five feet or something like that, in front of the headlights, they were shining right on it. It kind of looked back at us, but didn't break its stride, it was obviously getting out of there. But it wasn't really scared or anything, it was just walking. It was about eight feet tall. It had long arms. It wasn't fat like some of the pictures we've seen, it was kind of slender but its legs were really long, like they were disproportionate to the body and...that's what we saw and...my husband yelled something to the effect of, 'Holy shit, that's incredible, it's a sasquatch!' He yelled something like that.

"He wanted to leave the car and go after the thing, but we said no, we will drive around. We drove around the campground, we came to another truck that flashed its headlights at us so we stopped. There were four other people in the cab of the truck. We told them what we saw and they said that they saw something similar about twenty minutes earlier. But they were quite drunk, and only three of the four saw.

"We went back to our campsite and I would not get out of the car. I wanted to drive back to Calgary. Scott got back into the car with me but he was no help at all, he wanted to sleep. I stayed up all night and watched the trees, but when the deer came back I knew that it was all right and that the creature was gone. Darwin and Shannon stayed up all night by the fire. You know there are so many deer at this campsite. They will eat out of your hands. A couple of hours before we saw this thing I made the comment, 'Where have all the deer gone?' Like the deer had disappeared, there weren't any deer around, all right. In the morning when I saw that the deer had come back, I knew it was safe and whatever this thing was it had disappeared. The next morning, after much debate, we decided to report it to the warden, so we went to the warden's office. We had to wait for him, and when he showed up we told him our story."

Q: "What was his reaction?"

Susan: "His reaction was...he tried to convince us that we had seen a bear. That's the impression I got because he kept saying, 'Well, bears stand up on their hind legs, are you sure it wasn't a bear?' He was nice, but I think that he thought we had seen a bear."

Scott: "We were sitting by the picnic table playing cards and it was a fairly windy night because our cards blew off the table and we had to pick them up. It got late enough that I decided that I would like to go to bed. So we went to the car and got the toothbrushes, and then we started walking down this trail towards the washrooms, maybe 300 feet away around the corner. We heard this noise, a snort, as I described before, and Susan said, 'I'm scared, I don't want to go there.' I'm not sure of her exact words, but it was something like that. But I thought it was a deer because there are lots of deer around there, so I kept pulling her down the trail until she finally said, 'It's a bear!' and then everybody ran for the cars.

"But I didn't believe it was a bear. Susan gets scared every once in a while, so I was walking back, looking over my shoulder to see if I could see anything. I didn't see anything...that looked like a bear. It looked like a large human. It was really dark, black, and it was standing on the trail. I got back to the cars. We all got into the cars,

Susan and Darwin in the other car. They turned their headlights on, but they were facing in the opposite direction. I finally got my lights on, and this creature took three or four strides across the area in my lights, on the ridge, and it walked through the bush in that direction. And I got pretty excited about it because I thought, this must be what everybody says is a sasquatch. So I jumped out of my car and ran back to talk to Darwin and I wanted to go chase after it to see what it was. But my wife was not very keen on that idea, so we got into the car and drove around the campground.

"As we drove around, after awhile somebody came down and flashed their lights at us. A big truck, and there were three or four people in the front, and they said, 'Have you seen anything?' We said the same to them, and they described this creature that they had seen. There was one guy and two girls that went down to the bathroom, and as they were coming back they said they heard a noise and they looked and saw this big creature they thought was a sasquatch, and that is what they said. I don't think they saw it as well as we did."

Q: "You told the warden the next day?"

Scott: "Yes."

Q: "And he tried to convince you it was a bear?"

Scott: "Well, I was kind of reluctant to report it at first, because I thought what's going to come of this? Probably nothing. People are going to think we are cranks. But Susan and Shannon finally talked us into going to the warden and reporting it. We told him the story and like I said, I think he only half believed us that it was a sasquatch, and half believing that we saw a bear."

Q: "The written report you've given me, who wrote it?"

Scott: "Darwin."

Darwin: "As I said, we were playing cards to about 12:30. Myself and Shannon sat by the campfire and Scott and Susan joined us. After a short while Scott and Susan decided that they were going to get ready for bed, so they were going to go brush their teeth. They grabbed their brushes and began to walk down the path. As soon as they walked out of the campsite, we heard the first grunt or sound and I jumped and Scott says, 'It's just a deer,' and he grabbed Susan by the hand and continued down the path and this thing grunted again. This time Susan said something like, 'That's not a deer, it's a goddam bear!' and as soon as she said that I was up and so was Shannon. I believe it was Shannon who said, 'Let's move calmly towards the cars.' But it was a free-for-all for the cars, we just sprinted.

"In the confusion, I ended up in the same car as Susan, and Shannon and Scott ended up in the other car. Our car was facing in towards the bush, sort of away from where this thing was sighted. Susan was quite excited and she wanted to honk the horn. I stopped her from honking the horn and I looked back towards where we had seen the bear, so-called bear. And I saw a tall skinny shadow, and I thought what is this bear doing up on its back legs? Usually they come down off their hind legs if they are standing up. I turned on the headlights of our car, I couldn't see anything, so I rolled down my window and I yelled to Scott, to turn on his headlights, since his car was facing the other direction, and when he turned them on that's when all four of us saw this thing plain as day.

"It was leaving our campsite. Walking on an angle away from us, across the beam of the headlights, in the trees, it was about four or five feet into the bush. There's no mistake about what we saw. The thing was walking upright, it was about eight feet tall, it was swinging its arms, and it definitely had arms and legs. When I reported it to the warden, I referred to them as arms and legs, not hind legs. This thing was walking upright. After it was gone, Scott came running back to me, quite excited, and said, 'Did you see that thing?' I said, 'Yeah,' and we grabbed the girls and piled into his car and went looking for it. We didn't see it again.

We were circling around the campground, talking quite excitedly about what we had see. We saw another truck coming towards us. We flashed our high beams at them, and they came to a stop. We rolled down our window and asked them if they had seen anything strange this night. There were four people inside that truck, three of which said they had seen something big and black, when they were on the way to the washrooms. They were in site B, and I asked them for a time, and they said about 12:30, so just before us. So I'm convinced that they saw the same thing. Unfortunately they did not report it."

Author's note: The next day Darwin wrote out a written report for the park warden's office.

Shannon: "We had been playing cards, and we decided to sit around the fire for a while. Susan and Scott decided to brush their teeth, and they started walking down the path towards the washrooms. There was a noise and Susan jumped back. She was startled but Scott said, 'It's just a deer,' and started to drag her down the path. Then we heard her scream, 'That's no deer!' And we all looked up and you could see this tall dark outline against the tree. I thought that it was a bear too, and we all turned around and raced for the cars as fast as we could. We got into the cars, Scott and I in the one car, Susan and Darwin in the other. The cars had their back ends together, and our car is facing the bush. Darwin turned on the lights in his car, then we turned on the lights in our car. That's when we saw that it wasn't a bear, and we could see this thing from the side view walking through the bush, saw it take ten steps, it was really long. It was dark hair, and it had a gait almost like that of a human. Even when we turned the lights on it, it didn't speed up or try to hide from us. It was just as if it was going on its way and wasn't bothered by anybody, and by the time we realized what we were looking at, and sort of got our act together, we tried to follow it, but we never saw it again. It was a little while later we ran into other people who said they saw something similar."

After I interviewed the four witnesses, I phoned Park Warden Alan Dibb to get his thoughts on the entire matter. As far as he was concerned, these people must have seen a bear or been the victims of a hoax. He was impressed with the fact that no matter how hard he tried to convince them that it was something other than a sasquatch they saw, they were adamant and would not be dissuaded about what they had seen. He was willing to let the matter drop after talking to the four, but Darwin Gillies insisted on making out a written report. He told him that a written report wasn't necessary, but Darwin insisted.

Crandell Campground is often referred to by visitors as Crandell Lake Campground. This is wrong. The site's proper name is simply Crandell Campground. The camp is located on the west bank of Blakiston Brook, in Waterton Lakes National Park. To get there you must drive seven kilometers (4.5 miles) northwest on Redrock Canyon Road, until you see the turn off for Crandell Campground to the left. The area is wonderful mountain wilderness, rich in wildlife. Deer have grown used to people and come down out the mountains and will eat out of people's hands. Black and grizzly bear, as well as mountain lion, are often seen in the hills around Crandell Campground. I visited site C-3, to have a look at the area myself, take pictures and make measurements. Both Susan and Scott reported that when they first saw the creature on the trail it stood about ten to fifteen feet away. My own measurements of the area, based on where Susan and Scott said they were standing, to the point where the creature was about fifty-five to sixty feet. That's still darn close, and it is indeed possible that I was off slightly, since nobody made marks around their feet so the exact spot they were standing was indicated.

When the creature passed through the parked car headlights, Susan estimated the distance from themselves to the creature was thirty to thirty-five feet, Scott said forty-five feet, Darwin says twenty to thirty yards, as well as mentioning the creature was about five feet into the bush, Shannon reported twenty feet maybe. I think based again on my own measurements, the distance was about 100 feet. The creature was, if positioning is correct, slightly up hill from the witnesses when seen. Also, there is some dead ground between the creature and the witnesses.

When asked how long they saw the creature for, Susan said a minute, maybe two. Scott reports twenty seconds, Darwin three to four seconds, Shannon reported ten to twenty seconds. Interesting that Susan says the creature was in sight for a longer period of time than do the other three. It does seem she was the most frightened by this event, so maybe it just seemed longer to her.

All four reported that it was walking upright on two legs the whole time it was in sight. During their talk the next day with Warden Alan Dibb, he said things to them along the lines of, "Bears do stand up when begging for food, and they can walk on their hind legs." But all four were sure and they insisted it was not a bear. I think Darwin at the time was getting a little annoyed at the warden's continuing suggestion of a bear being the creature they had seen. In his written report he says, "We, the four of us, are convinced it was not a bear." Also he ended his report with the statement, "It wasn't a bear, O.K!"

All four reported that the creature was between seven and eight feet tall, most likely eight feet. I feel it should be noted that when all four were watching it from the cars, it is possible the creature was slightly uphill from them. Whether or not this made the creature appear taller than it actually was has to be considered. However, they were close enough in my opinion to be pretty accurate.

All four couldn't tell the gender of the creature that night. However both Scott and Shannon guessed that it was probably male due to the fact it seemed to lack female mammary glands (breasts); however, they couldn't say with any certainty.

When asked the creature's color Susan reported black or dark brown, Scott reported dark brown or black, Darwin reported black. Shannon couldn't really say, but described it as a dark color. All four were referring to the creature's hair color. None noticed the color of the creature's skin.

None noticed any facial features, Susan just said that it was a flat face.

All four reported that the creature was rather thin in stature, but huge in general size.

Three reported that the creature's arms were very long and hair covered. Scott didn't notice the arms all that much.

Both Susan and Shannon describe the noise coming from the creature as a grunting sound, Scott says it was a snorting noise much like a bull makes, while Darwin says "a snort or a grunting sound." None felt the creature was trying to be aggressive, and the noise was just a way of letting them know it was there, and not to come closer

They did look for footprints the next day, but none were found. I found the area to be typical of high mountain forest terrain. The path to the washrooms is hard-packed gravel, so unless the subject walked into a soft muddy spot somewhere, I'm not surprised that no footprints were found. As far as I know the search for prints the next day was in the area of sites C-3 and C-4. As far as I know there was no searching down by Blakiston Brook.

Who were the other people in the truck who also claimed to have seen the creature? All my attempts at locating these people have failed. They did not report their encounter the next day. Also, the warden's office did not try to locate them the next day. Darwin mentioned that the other four were staying in camping area B, which is located just southwest of area C, where our four witnesses were staying. Could these four have pulled off a good hoax and later hung around to see if their joke had been successful?

One detail about the creature itself bothers me. Scott said that it walked with very little bending of the knee. And he was very impressed by this. Maybe so, but it doesn't fit with most other descriptions of the sasquatch when it walks. If anything it has been reported that a sasquatch walking bends the knee more than a human, especially when the creature is leaving in a hurry. The subject in the Patterson/Gimlin film seems to lift the knee higher than a man would. Could the lack of knee movement here be because of a loose-fitting costume? All four witnesses here are mature professional people, who watched it at very close range. Darwin said himself at first he thought it was a joke, but quickly changed his mind. The fact the other four people in the truck didn't report this to the warden's office isn't surprising. Most people who see a sasquatch don't. Susan, Scott, Darwin and Shannon are the ones who did the rare thing, they reported it. I think what we really have here is an incident which really has seven witnesses rather than four.

It has been ten years now since that night in 1988. Susan Adams and Scott Stoness have since divorced and lead separate lives. Darwin and Shannon on the other hand have since married, have three children and are very happy. Scott Stoness now seems to be trying to forget what happened in 1988, and is carrying on with his life. The last time I spoke to him he seemed tired of talking about it and told me, "I don't know what we saw!" Susan Adams is still a school teacher and is quite open with her students with what happened that night. She is often asked by other teachers to talk with their students whenever they are discussing Canadian mysteries, and the subject of the sasquatch is being talked about.

In April, 1997, I was asked by Mr. Lorne Monro to come to his class at Mackenzie Lake Elementary School in Calgary to talk with his students about the sasquatch. I often get such requests, and do so whenever I can. As I was sitting in the teachers' cafeteria before his class was to begin, Susan walked in. She too was asked to speak to the students. I watched her as she told the transfixed young men and women the same story she told me nine years earlier. We talked for a while, and I found out she is often asked to relate her story in class. She does so without any hesitation.

Shannon Gillies is also a teacher, and also talks to students about her sasquatch encounter in 1988. Her husband Darwin still works as an engineer, and he and Shannon have often returned to Crandell Campground. They told me that it is almost a ritual for them. Of course, sasquatch seems to be the favored topic around the fire. As far as I know neither Susan nor Scott have ever been back to Crandell Campground. Susan tells me she is still a little scared in the woods at night.

Park Warden Alan Dibb no longer works at Waterton Lake National Park. I've been told he works up in the Yukon now.

The last time I was at the campground was in May, 1997. I was trying to take more measurements for a paper I was to give in Vancouver at the International Sasquatch Symposium in June. The wardens who work there now have all heard of the sasquatch incident in Crandell Campground, though they all seem rather skeptical about the whole thing. They told me that nothing has been reported there since, at least nothing has been brought to their attention.

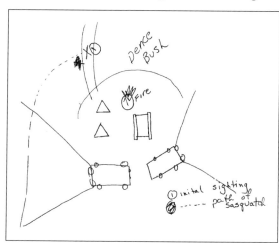

Diagram and map of creature's course at Site C-3, Crandell Campground. Drawn for me by Susan Ray Adams, when I first interviewed her in 1988.

245

Above: Site C-3, Crandell Campground, Waterton Lakes National Park. It was here on May 23, 1988, Susan Ray Adams and Scott Stoness, as well as friends Darwin Gillies and his then girlfriend, now his wife, Shannon Senkow encountered a creature which fits the description of the sasquatch.

Left: The pathway behind site C-3 where Susan and Scott nearly walked into the creature as they were going to the public washrooms to brush their teeth.

Below: The small wooded area which was illuminated by the car headlights. All four witnesses watched the creature cross right to left at close range. All four are adamant that this creature could only have been a sasquatch.

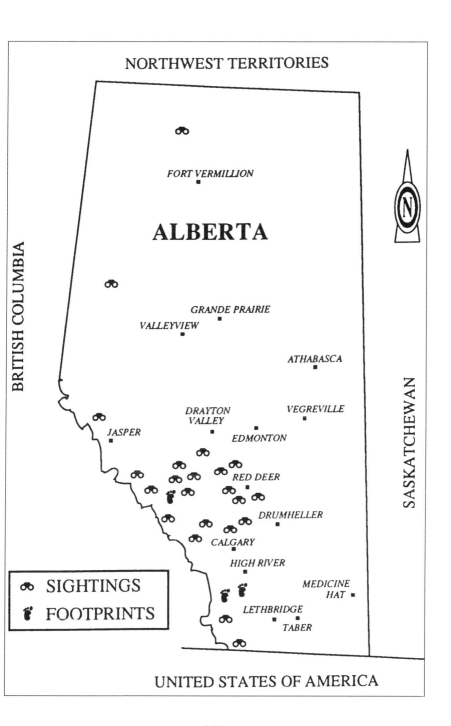

NORTHWEST TERRITORIES

ALBERTA

BRITISH COLUMBIA

SASKATCHEWAN

FORT VERMILLION

GRANDE PRAIRIE

VALLEYVIEW

ATHABASCA

DRAYTON
VALLEY

VEGREVILLE

JASPER

EDMONTON

RED DEER

DRUMHELLER

CALGARY

HIGH RIVER

MEDICINE
HAT

LETHBRIDGE

TABER

SIGHTINGS
FOOTPRINTS

UNITED STATES OF AMERICA

Alberta
Statistics

As in chapter six, these statistics are based solely on information in my files. They are not intended to make any general statements as to the sasquatch's habits, or when and where someone is likely to run across one of these creatures. Again if they did reflect such accurate information about this creature, we probably would have brought one in by now. These Alberta statistics do differ at times when compared to British Columbia. I must point out that I have many more reports in total from British Columbia than Alberta, and the difference could simply be due to this fact. These statistics are based on the number of reports I had at the time the graphs were made—thirty-four reports. At the time of writing, I now have forty-four reports of encounters in Alberta. When my book The Sasquatch in Alberta came out in 1990, I had twenty-three reports on file. As you can see, the number increases every year.

In Alberta, most reports occur during the summer season from May 24 to September 24; seventeen incidents were reported during the summer months (50 percent of the total). I have nine reports from the fall (27 percent), followed by five spring reports (15 percent). I have only three reports during the winter (7 percent). These figures seem to be approximately the same as British Columbia.

Despite the belief that the sasquatch is a nocturnal creature, my statistics show the opposite. In Alberta, twenty-six reports occurred during daylight hours (79 percent). Another seven sightings occurred at night (21 percent). Again the same percentage almost as in my British Columbia files.

As in British Columbia, the gender or sex of the creature has for the most part gone unnoticed. Twenty-five reports give no indication to the sex of the creature observed (76 percent). Five reports indicate the creature was a male (15 percent). Three reports said the creature was female (9 percent). All reports of female creatures are due to the witness seeing female mammary glands (breasts). The male reports are due to the lack of mammary glands. Not one report from Alberta makes notice of male genitalia.

The most common hair color reported in Alberta is black, that is fourteen reports (43 percent). This differs from British Columbia, where the animal has most often been reported to have brown hair. There were six reports of brown hair (18 percent). Reddish brown coloring has been described in five reports (15 percent). I have another four reports where the witness said the creature was either blackish brown or brownish black (12 percent). One report of a black creature with gray tips (3 percent), and one report of a gray-colored creature (3 percent). Two reports make no mention of the creature's hair color (6 percent).

As you have read earlier in this book, Alberta has had more than its share of extremely tall creature reports. However, these cases are by far the minority. The most common height report in Alberta is eight feet. There have been seven reports of eight-foot creatures (23 percent). This differs again with British Columbia where the most common height is seven feet. There have been six reports of seven-foot creatures (20 percent) and three reports of seven and a half to eight feet (10 percent). In addition, there are three reports of six and a half to seven-foot creatures (10 percent). I have one report of a sasquatch that stood between ten and twelve feet (3 percent) and two more reports of twelve-foot creatures (7 percent). It must be pointed out, it is hard to be accurate when suddenly confronted by a creature that you always assumed didn't exist. Many of these reports consisted of witnesses making guesses at the height of the creature they were seeing. I assume they could be off by a foot here or there.

As I said in chapter six, when more than one witness sees a crime being committed it is known as corroboration. However, in the world of science, it is most likely to be referred to as mass hallucination as far as the sasquatch is concerned. In Alberta, one incident had five witnesses (3 percent). Four incidents had four witnesses (12 percent). Four more situations had three witnesses (12 percent). Eight events had two witnesses (24 percent). The majority of incidents were one person claiming to have seen a sasquatch— sixteen incidents (49 percent). If one was to add all the incidents with more than one witness together, they would total seventeen events—one more than single-witness sightings. This again differs from British Columbia, where single-witness cases outnumber all other cases.

Again this chapter was not intended to give any specific impression of sasquatch living habits. The information was based solely on data taken from my own files. I just thought some of the differences between Alberta reports and British Columbia reports were interesting. In another ten years or so, when I update the graphs, the outcome may be completely different.

Personal Thoughts

In this book I've concentrated on eyewitness testimony from British Columbia and Alberta, from the east slopes of the Rocky Mountains west to the Pacific Coast. This has been my main area for research in the last twenty years. I do have reports from the Pacific Northwest of the U.S. as well, but I do not have the chance to look into them too often anymore. And if the Canadian dollar continues its downward slide, it doesn't look like I will get the chance in the near future either. However, I do have contacts in all the western states of the U.S. If anything is reported to me, I pass the information to appropriate American colleagues and let them follow up on it. In return, they do the same when they hear of something happening in eastern B.C. or Alberta.

Until a few short years ago, I was always skeptical of reports from central Canada, as well as the eastern parts of the United States. It always seemed to me most lunatic fringe reports came from the eastern part of the continent, such as reports of three-toed and six-toed footprints, sasquatch with illuminating red eyes, creatures connected with UFOs and other strange, unexplained things. Whereas in the western parts of the two countries, reports always seemed to be of a reclusive, big, bipedal creature that was nothing more than an unclassified animal. However, I have since found that we in the west have our share of fringe types as well.

Sasquatch enthusiast Scott Herriott of California, is a very funny stand-up comedian who produced a very funny video entitled *Journey Towards Squatchdom*, in which he visited some of the most well-known western lunatic fringers, as well as a few serious researchers. The end result was one of the most funny and entertaining videos I've seen in a long time. Scott does have a serious side as well and was featured in an A&E documentary in which a video was shot supposedly showing a real sasquatch hiding in some thick bush in Northern California. Whether or not there was a real sasquatch there remains to be seen. Something was moving in those trees; however, you never get a very good look at it. Scott sent me a copy of that video and I really cannot say that a sasquatch was hiding in those trees. It could have been some common animal. However, if it was a sasquatch, it is an excellent example of how well these creatures can hide themselves when disturbed. But as evidence, it doesn't really help much.

Some serious researchers in the east have me thinking twice about the possibility of this creature living in remote forest regions there as well. Ontario is Canada's largest province with thousands of miles of dense, humid forest, more square miles of forest than in B.C. as a matter of fact. I had not heard of any reports of sasquatch in Ontario until I moved west. On the border region with Manitoba, there have been many reports. The town of Cobalt, Ontario, had a creature in their area for years. Locals gave it the name Old Yellow Top because it had a streak of light-colored hair on the top of its head. It was around from the late 1950s through the 1960s. The last reported sighting of Old Yellow Top occurred in 1970. There have been none since, at least none that I've heard of.

Author searching for footprints along Spray Lakes in Kananaskis country, Alberta, early spring 1994.

Photo: T. Steenburg, 1994.

The scientific community here in North America has for the most part dismissed any notion that sasquatch may indeed exist. There have been noteworthy exceptions to this, such as professors Markotic, Krantz, Napier, Meldrum, Koffmann and others, but for the most part nothing has changed in the last forty years or so.

Science needs hard evidence, not eyewitness testimony. What works for the legal system, does not work for science. Even films and photographs are considered soft evidence as far as most scientific organizations are concerned, and I understand this position. If science were to declare a new species based on only oral testimony, then unicorns, mermaids, vampires, werewolves, ferries, goblins and many other mythical creatures would be on the endangered species list.

The scientific community in Russia though seems more willing to consider the possibility that the sasquatch does indeed exist. Professors like the late Boris Porshnev, and Dmitri Bayanov and Marie-Jeanne Koffmann, have declared publicly that as far as they are concerned the creature does indeed exist. In 1971 Rene Dahinden, who was frustrated with the lack of interest in the Patterson/Gimlin film, took the film to the Darwin Museum in Moscow, so the Russians could have a look at it. The response was a lot more receptive than any North American organization. As far as they were concerned the Patterson/Gimlin film does indeed show a living creature. However, cold war politics and attitudes prevented the Russians from making their finding s known for the most part here in North America. Their findings, now that the cold was is over, are really just beginning to be well received here. I don't think they will change any of their North American colleagues' minds; however, it is a start.

Unlike the Russians, most North American scientific organizations have decreed that the mystery of the sasquatch will continue until hard physical evidence is brought in and laid out on a table. What do they mean by "hard physical evidence?" Quite simply, they need a body, or at least significant portion of a body. This could mean skeletal remains. However, there have been stories in the past of large jawbones, femurs and pelvic bones being found in the woods of northwest America. But the bones were almost

Still 352, from the Patterson/Gimlin film, shot at Bluff Creek, Northern California, on October 20, 1967. This thirty-year-old film still remains the best evidence to date of the existence of the sasquatch.

Photo © Rene Dahinden, 1968.

always lost, put away somewhere or identified as belonging to some common animal. It wouldn't surprise me at all if the day comes and the skeletal remains of a sasquatch are laid out on a table for all the world to see. And instead of being brought in from the woods by some hunter or researcher, they may be laid out by some anthropology student who stumbled on them in a long-forgotten museum box or drawer. Until that happens, the onus is on us nonlunatic fringe researchers to provide the world of science with the evidence they need. What this means in straightforward, politically incorrect language is this—a sasquatch has to be shot and killed, and brought in. Nothing less will satisfy the scientific world that this creature lives, eats, reproduces and is alive and well in our remote wilderness regions.

Many people in the sasquatch field are against shooting for moral reasons, and I do understand how they feel. Before 1989 I too was against this. I thought a good, clear photograph or video would do. However, I now realize that if I did come back from one of my trips in the back country with a good video of a sasquatch, most scientists would simply accuse me of fakery. They would contend that, even though I do not have the technical skill to do such a thing, I must have faked it, because sasquatch do not exist. We would still be hearing the same question. Do these creatures really exist or not?

Others worry that shooting a sasquatch could endanger the species, and if the creature is indeed as rare as some think, it wouldn't be a good idea to shoot a bunch of them. I don't believe the population, based on how widespread the reports are, is so low that killing one or two creatures could terminate the species. If the population is that low, then they are doomed anyway. Killing a sasquatch would prove its existence. This is a fact. Photographing or videotaping a sasquatch probably would not prove it. Those against shooting are reacting with emotion, not common sense. They are reacting with their heart, rather than with their head. I don't mean to criticize them for this, for in many ways I feel the same. I don't even know whether I could pull the trigger if I had one in my rifle sites. The simple fact is that only hard physical evidence will prove the existence of this creature. It has been stated time and time again.

I sometimes wonder if those who are against the shooting of a sasquatch really want to see this mystery solved at all. After all, if the creature is proven to be real, then the puzzle will be over. One of the great questions of cryptozoology will suddenly be reduced to the realm of everyday zoology. What will we do then—move to Loch Ness? I don't care if I happen to be the one to prove its existence. I would love it if I was, but I just want to find out one way or the other. My friend the late Vladimir Markotic once said to me during one of our late-night debates at his cabin, "Tom why should we kill a sasquatch to prove its existence anyway? If it does exist, it will be a major scientific discovery. If the creature does not exist then it has to be researched anyway, because it is an important part of our folklore." He gave me new food for thought there, but I still maintain that until the creature is proven to exist, then what they are, where they came from, and what their future holds is irrelevant. We must prove their existence before anybody will really care about all the other aspects of their being.

Do I personally believe this creature exists? The answer to this question is yes. All the evidence so far in my opinion may be circumstantial, but it is convincingly so. So one of the greatest mysteries of the twentieth century looks as if it will still be one of the greatest mysteries of the twenty-first century. Until a body is produced, we will be no closer to answering this question. So I will close with this thought. We have to have a body. Nothing else will do. If I'm wrong, and a photograph or video does prove the existence of this creature, then I will gladly admit I was wrong. And I will praise the photographer with accomplishing the unaccomplishable (assuming there is such a word). Until then, nothing less than hard physical evidence will suffice. We may not like it, but that is reality.

Anyone who has information concerning sasquatch
is invited to contact me:

Thomas Steenburg,
Calgary, Alberta